INTERNATIONAL ENERGY AGENCY

AF886202

Toward a SUSTAINABLE ENERGY FUTURE

OECD

INTERNATIONAL ENERGY AGENCY

9, rue de la Fédération,
75739 Paris Cedex 15, France

The International Energy Agency (IEA) is an autonomous body which was established in November 1974 within the framework of the Organisation for Economic Co-operation and Development (OECD) to implement an international energy programme.

It carries out a comprehensive programme of energy co-operation among twenty-five* of the OECD's thirty Member countries.

The basic aims of the IEA are:

- To maintain and improve systems for coping with oil supply disruptions;
- To promote rational energy policies in a global context through co-operative relations with non-member countries, industry and international organisations;
- To operate a permanent information system on the international oil market;
- To improve the world's energy supply and demand structure by developing alternative energy sources and increasing the efficiency of energy use;
- To assist in the integration of environmental and energy policies.

**IEA Member countries : Australia, Austria, Belgium, Canada, the Czech Republic, Denmark, Finland, France, Germany, Greece, Hungary, Ireland, Italy, Japan, Luxembourg, the Netherlands, New Zealand, Norway, Portugal, Spain, Sweden, Switzerland, Turkey, the United Kingdom, the United States. The European Commission also takes part in the work of the IEA.*

ORGANISATION FOR ECONOMIC CO-OPERATION AND DEVELOPMENT

Pursuant to Article 1 of the Convention signed in Paris on 14th December 1960, and which came into force on 30th September 1961, the Organisation for Economic Co-operation and Development (OECD) shall promote policies designed:

- To achieve the highest sustainable economic growth and employment and a rising standard of living in Member countries, while maintaining financial stability, and thus to contribute to the development of the world economy;
- To contribute to sound economic expansion in Member as well as non-member countries in the process of economic development; and
- To contribute to the expansion of world trade on a multilateral, non-discriminatory basis in accordance with international obligations.

The original Member countries of the OECD are Austria, Belgium, Canada, Denmark, France, Germany, Greece, Iceland, Ireland, Italy, Luxembourg, the Netherlands, Norway, Portugal, Spain, Sweden, Switzerland, Turkey, the United Kingdom and the United States. The following countries became Members subsequently through accession at the dates indicated hereafter: Japan (28th April 1964), Finland (28th January 1969), Australia (7th June 1971), New Zealand (29th May 1973), Mexico (18th May 1994), the Czech Republic (21st December 1995), Hungary (7th May 1996), Poland (22nd November 1996), the Republic of Korea (12th December 1996) and Slovakia (28th September 2000). The Commission of the European Communities takes part in the work of the OECD (Article 13 of the OECD Convention).

© OECD/IEA, 2001

Applications for permission to reproduce or translate all or part of this publication should be made to:
Head of Publications Service, OECD
2, rue André-Pascal, 75775 Paris cedex 16, France.

FOREWORD

Energy has deep and broad relationships with each of the three dimensions of sustainable development – the economy, the environment and social welfare. While certain forms of energy production and consumption can diminish environmental sustainability, energy is crucial for economic development. Energy services also meet basic needs such as food and shelter; they contribute to social development by improving education and public health. In working toward sustainable development, the manner in which we produce and consume energy is therefore of crucial importance. This book is a first overview of policies toward sustainability in the energy sector.

Two criteria have been identified for such policies. First, they must strike a balance among the three dimensions of sustainable development – the economic, the environmental and the social one. Second, they must reduce our exposure to large-scale risks and improve the resilience of the energy system through active risk management and diversification. These two criteria will have to prove themselves in the face of several potential challenges to sustainable development:

- The growing importance of developing countries in global energy consumption will have important consequences for global supply security and greenhouse gas emissions. In addition, nearly 2 billion people lack access to electricity.

- Climate change is a major issue for industrialised countries, while developing countries need to address local and regional pollutants from fossil fuels.

- The security of energy supply is increasingly at risk as the dependency of OECD countries on a dwindling number of oil and gas suppliers grows.

- The institutional environment for energy policy-making is changing fast. Just as market reform and privatisation are decentralising decision-making power, more and more issues require regional or global co-ordination.

- Improvements in energy efficiency and reductions in the cost of renewable energy sources – that hold great potential for progress for sustainable development – require careful nurturing and the right policy frameworks.

A number of initiatives demonstrate that policymakers are responding to these challenges. Both the Ninth Session of the Commission on Sustainable Development in New York in April 2001 and the Meeting of IEA Energy Ministers in Paris in May 2001 have the relationship between energy and sustainable development as their principal theme. The OECD horizontal project on Sustainable Development provides a reference for a large number of policy efforts aimed at increasing sustainability. Following the principles in its *Shared Goals* of 1993, the IEA seeks to contribute to these efforts by providing information and policy advice in the energy sector.

In the context of this work we found it useful to have a working definition of "sustainable development" in the context of energy policy-making. In this book the term will be taken to mean *'development that lasts' and that is supported by an economically profitable, socially responsive and environmentally responsible energy sector with a global, long-term vision.*

This work is published under my authority as Executive Director of the IEA and does not necessarily reflect the views or policies of IEA Member countries.

Robert Priddle
Executive Director

ACKNOWLEDGEMENTS

Toward a Sustainable Energy Future is the result of a broad horizontal effort throughout the International Energy Agency (IEA) that involved all four of the IEA's directorates. The study was carried out in the Office of Long-Term Co-operation and Policy Analysis under the leadership of Olivier Appert, Director. Jonathan Pershing, head of the Energy and Environment Division, provided the initial impetus for the publication and assured day-to-day managerial oversight. Jan Horst Keppler, Principal Administrator in the Long-Term Office, wrote the introductory chapters and co-ordinated the different contributions.

Subject chapters were contributed by Kristine Kuolt (Emergency Planning and Preparedness Division), Carmen Difiglio (Head, Energy Technology Policy Division), Rick Sellers (Energy Technology Collaboration Division), David Wallace (Energy Technology Collaboration Division), Richard Baron (Energy and Environment Division), Lew Fulton and Michael Landwehr (Energy Technology Policy Division) and Xavier Chen (Non-Member Countries Division). Mel Kliman and Chris Henze, independent consultants, edited the final manuscript. Finally, the study benefited from discussions on energy and sustainable development with a number of experts at the International Energy Agency and the OECD including Shimon Awerbuch, Marco Mira d'Ercole, Trevor Morgan, Cedric Philibert, Ronald Steenblik, Kristi Varangu and Ann Vourc'h.

TABLE OF CONTENTS

LIST OF BOXES

LIST OF FIGURES

LIST OF TABLES

INTRODUCTION AND BASIC CONCEPTS

Introduction

The idea that sustainable development should be a key objective in government policy is becoming a reality. When the Brundtland Report was published in 1987, it set in motion a major reconsideration of economic policy. Its central prescription – *development that meets the needs of the present without compromising the ability of future generations to meet their own needs* – has slowly evolved from a slogan for environmentalists into a guiding principle for public policy.

The implications of sustainable development as a policy objective are inherently complex, all the more so because its advocates insist that it should be interpreted broadly. Development policy affects the economy, the environment and the social framework in which we live. Development should be sustainable in all three of these dimensions. Substantial efforts are underway to analyse and interpret sustainable development as a practical guide to action. The relation of sustainability to established policy tools – such as targets for economic growth, national accounting systems, criteria for environmental protection and social welfare – needs to be better understood.

This process is proceeding on a sectoral basis. Energy is an especially important sector in this context. Energy affects all of the dimensions of sustainability. Sustainable development therefore presents the International Energy Agency with a major challenge, and the Agency has placed a high priority on responding to it. Numerous activities in the IEA's governing body, working committees and Secretariat are currently focused on understanding the implications of this issue and on supporting Member countries in their efforts to deal with it.

The purpose of this book is to contribute to provide context for understanding sustainable development in the energy sector. In this chapter, we discuss some background aspects of sustainable development and its analysis. This prepares the way for the exploration in the next chapter of a number of important concrete issues as they apply to the energy sector. Chapters 3-9 then deal in more detail with seven areas of energy policy making that have important implications for sustainable development. These are: energy supply, security, market reform, improving energy efficiency, renewable energies, sustainable transport, flexibility mechanisms for greenhouse gas reductions and non-Member countries. Conclusions and recommendations are set out in a final chapter.

We aspire to provide a sense of strategic direction in regard to policy decisions in the energy sector. We do not dodge the difficult questions nor do we offer *passe-partout* responses that ignore context and changing circumstances.

Aspects of Sustainable Development

An important element of the broad context of the discussion about sustainable development is the argument that it should be approached in all of its primary dimensions – economic, environmental and social. In the past, development issues have tended to be considered more narrowly, mainly in their economic dimensions. The impetus for the sustainable development movement is in part a reaction to that way of thinking.

A recent report from the OECD's current project on sustainable development describes its three dimensions as follows:[1]

Economic *sustainability encompasses the requirements for strong and durable economic growth, such as preserving financial stability and a low and stable inflationary environment.* Environmental *sustainability focuses on the stability of biological and physical systems and on preserving access to a healthy environment.* Social *sustainability emphasises the importance of well functioning labour markets and high employment, of adaptability to major demographic changes, of stability in social and cultural systems, of equity and of democratic*

1. OECD (2001), Analytic Report on Sustainable Development: Chapter 2: Key Features and Principles, *p. 4.*

participation in decision making. These requirements are recognised as distinct from, and as important as, economic efficiency. Sustainable development emphasises the links among these three dimensions, their complementarities, and the need for balancing them when conflicts arise.

The discussion in this book takes account of all three dimensions of sustainability as they relate to the energy sector.

From that starting point, this section explores several aspects of the sustainability concept at a general level with a view to special implications for the energy sector: the orientation of sustainable development toward the future, its global character, the existence of synergies among the different goals associated with sustainability and the scope for substitution among the three primary dimensions of sustainability.

Orientation to the Future

Development that meets the needs of the present without compromising the ability of future generations to meet their own needs – or as David Pearce characterises it, "development that lasts"[2] – is strongly oriented toward the future. Policies to achieve it must be judged in terms of what they will accomplish over time, rather than by the achievement of a particular goal by a specified date, and in terms of how they respond to expected and unexpected change. Systems need to be robust, capable of evolving along a sustainable development path without being derailed by unforeseen events.

Such an orientation immediately raises the question of the relevant timeframe. It is correct but not very helpful to say that all timeframes – short-, medium- and long-term – are important. But how long are these timeframes? The very long run – 100 years and more – is important in environmental issues, such as climate change

2. The full citation reads "Sustainable development is development that lasts, and which is not therefore threatened by actions we take now and which will have their major consequence in the future." In "The Economics of Sustainable Development" by D.W. Pearce, G.D. Atkinson and W.R. Dubourg (1994).

or nuclear waste disposal. But economic policy is usually aimed at a five-to-ten year time horizon. Polices on social issues are frequently undertaken in a still less-clearly defined timeframe. In the social area, sudden shifts have to be taken account of; so the short run is highly relevant. Issues such as security for the elderly instead require a longer view in both social and economic terms.

Concern over the long term is particularly relevant for the energy sector since the effects of energy production and consumption can be felt for many generations to come. The physical capital that energy production typically mobilises lasts for several decades, sharply reducing the scope for swift reactions to changing circumstances. Climate change effects may reach over several hundred years. Nuclear waste can radiate for several thousand years.

Over the short and medium run, an important question for the energy sector is whether consumption and production patterns are sustainable. If present trends continue, including the heavy reliance on fossil fuels, risks will build up not only in the environmental but also in the economic dimension. The growing dependency of many

Figure 1

World Total Primary Energy Supply by Fuel

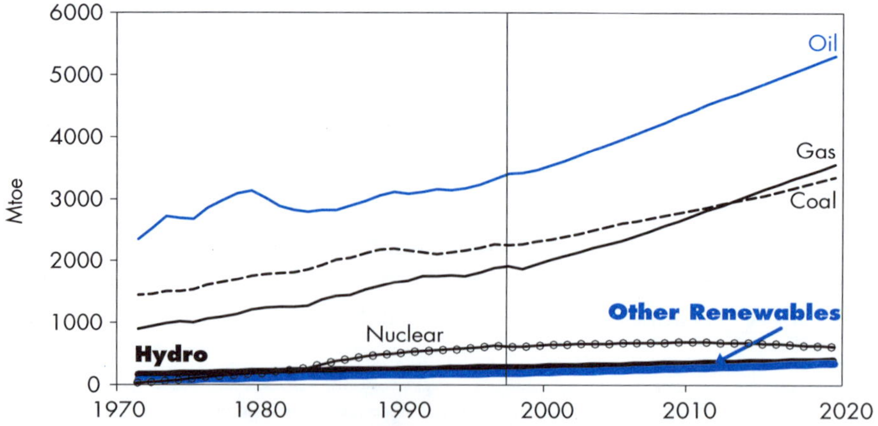

Note: IEA (2000), World Energy Outlook 2000, *p. 48.*

IEA Member countries on oil and gas imports will threaten supply security. In the energy sector, the future orientation of policymaking is not an abstract principle but a concrete necessity.

Globalisation and other Structural Changes

Sustainable development has to be considered not only through time, but also increasingly through space. Globalisation and the increasing interconnectedness between countries and regions are important aspects of several sustainability-related energy issues. Those include world oil markets, climate change and the challenge of providing access to electricity for nearly 2 billion people currently without it. The *World Energy Assessment* sounds alarm in this respect:

The current energy system is not sufficiently reliable or affordable to support widespread economic growth. The productivity of one-third of the world's people is compromised by lack of access to commercial energy, and perhaps another third suffer economic hardship and insecurity due to unreliable energy supplies; wide disparities in access to affordable commercial energy and energy services are inequitable, run counter to the concept of human development, and threaten social stability[3].

For the IEA and its Member countries, co-operation with developing countries is of great importance[4]. The growing importance of non-Member countries in global energy markets is displayed in the graph below.

Globalisation is becoming an ever more inescapable reality in energy matters. It requires special attention in an organisation such as the IEA because our membership has historically been restricted. Ten years ago, non-Member country issues were not yet an important topic in policy-related studies of this sort, except in

3. *UNDP, UNDESA and WEC (2000),* World Energy Assessment: Energy and the Challenge of Sustainability, *p. 5f.*

4. *There are 25 Member countries in the IEA: Australia, Austria, Belgium, Canada, the Czech Republic, Denmark, Finland, France, Germany, Greece, Hungary, Ireland, Italy, Japan, Luxembourg, the Netherlands, New Zealand, Norway, Portugal, Spain, Sweden, Switzerland, Turkey, the United Kingdom and the United States. Membership in the OECD is made up of these same countries and, in addition, of Iceland, Mexico, Poland and the Republic of Korea. The European Commission also takes part in the work of the IEA and the OECD.*

regard to the supply outlook in non-Member producing countries. Ten or 20 years ahead, the very distinction between the two groups may be obsolete.

Globalisation also poses the question of the scale on which sustainable development in the energy sector should be pursued: that of the energy sector itself? the national level? or the regional and global levels? Again the obvious answer – all of them – is not very helpful. Energy industries have long tended to have a global outlook. From the viewpoint of policy, the national level is still of primary importance. However, increased regional co-operation, as in the European Union, and international and global co-ordination among governments are becoming increasingly important.

Figure 2

World Primary Energy Demand by Region

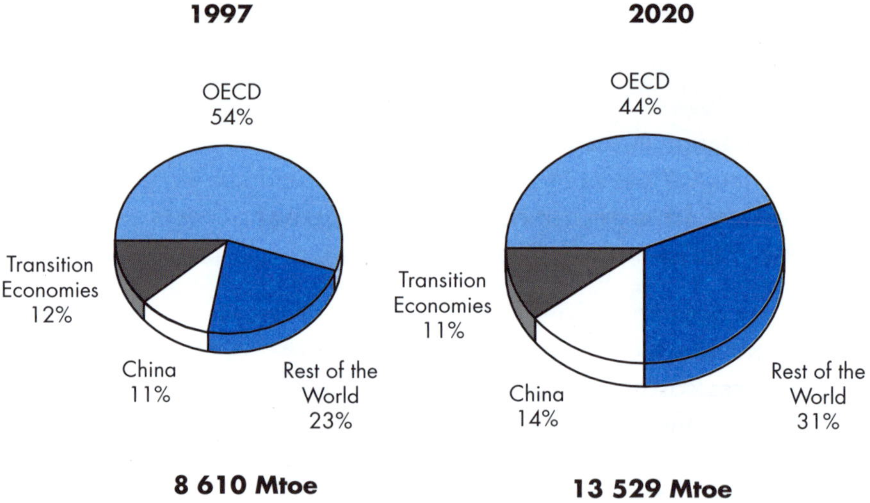

Note: IEA (2000), World Energy Outlook 2000, *p. 52.*

Ever-increasing globalisation, however, is not the only structural change sweeping through the energy sector. Regulatory change, new technological developments, an increased awareness of environmental issues and the availability of new information and communication technologies all offer challenges and opportunities for the energy sector. These structural changes can interact and produce new and unforeseen results. An impressive example has been the interaction of energy-market liberalisation, a mature infrastructure for natural gas and the advent of the combined cycle gas turbine. This convergence has allowed several OECD countries to combine economic efficiency gains and lower energy prices with reductions in greenhouse-gas emissions.

The impact of information and communication technology raises particularly interesting questions. Developments such as "goods travelling instead of people" in electronic commerce and the increased availability of choice for consumers of power and energy services are two good examples. Each could affect sustainability in the energy sector either positively or negatively. Another set of key issues involves new environmentally-friendly energy technologies: fluorescent light-bulbs, energy-efficient household appliances, decentralised renewable energy services, clean coal-technologies and advanced fuel cells. Care must be applied to advancing these technologies without violating the commitment to market-oriented organisation in the energy sector.

Policy Integration and the Quest for "Win-win" Opportunities

A key question on the way toward sustainable development is the extent to which the realisation of goals in the different dimensions requires costly trade-offs. Or can all three dimensions be pursued with a common policy?

To a degree this issue is easier dealt with in the energy sector due to the fact that energy use is not an end in itself, but rather a means to an end. Energy services – transport, travel, lighting, heating and cooling – not energy *per se* are the real outputs of the energy sector. The same ends can be provided by different means – different fuels, transformation systems and end-use technologies. As each fuel and each technology has different economic, environmental and social effects, choosing and mixing the different components of energy services allow policy-makers a good deal of flexibility.

In some cases, structural change, especially technological progress, permits progress on all three fronts at once. For instance, the advent of the combined-cycle gas turbine has allowed several OECD countries to combine economic efficiency gains with reductions in greenhouse gas emissions. Shifts in the economy toward knowledge-intensive services may allow us to reap similar benefits. The advent of new information and communication technologies may help us to market energy services with particular attributes, such as "green" electricity.

It is important to use the additional latitude offered by new technological and structural developments to move forward in all dimensions of development. Unfortunately, such "win-win" opportunities, which allow simultaneous improvements in the economic, environmental and social dimensions often take time to develop, and they occur in limited amounts. Timing and the setting of priorities are of great importance in the quest for win-win opportunities. The longer the timeframe, the greater the opportunities. Yet, it is crucial that policy priorities such as the struggle against climate change remain firmly in place over the full time period for which this freedom of manoeuvre applies. Win-win opportunities must not be dissipated under the stress of competing demands.

In the long run, the quest for win-win opportunities is made easier by the intrinsic linkage of the different dimensions of sustainable development. Something that looks like a trade-off in the short run may reveal itself over time as presenting advantages in two or even three dimensions at once. Complementarities among the different dimensions abound. Economies can only grow if they are not threatened by environmental catastrophe or social unrest. Environmental quality can only be protected if basic economic needs are fulfilled and individuals take responsibility for public goods. Finally, social development depends on economic growth and a healthy environment. Integrating considerations of environmental quality and social stability into decision-making in the energy sector not only involves costs in that sector, but also contributes to its long-run success.

Maintaining and improving environmental quality along with economic growth involves improving energy efficiency and switching to environmentally less harmful fuels, such as renewable energies which are discussed in Chapter 5 and 6 respectively. Balancing economic growth with social progress requires attention to issues such as consumer access and the global equilibrium of energy production and consumption discussed in the context of non-Member countries in Chapter 9.

"Weak" and "Strong" Sustainability

Notwithstanding the existence of win-win opportunities, progress in one dimension of sustainable development often implies losses in another. In these situations trade-offs need to be made. For example, reducing the cost of energy may provide wider access to it, but will result in more consumption and hence greater negative impacts on the environment. If energy prices are raised to combat environmental threats, energy services may be out of the reach of the consumers who need it most. In some cases, compensations can be provided to make the trade-off acceptable. The social hardship due to higher energy prices might, for example, be offset by direct payments to the needy. In other cases, however, such trade-offs are scarcely possible. It is unrealistic, for example, to argue that higher income growth could offset the catastrophic consequences of climate change.

Discussion on this point has resulted in a distinction between "weak" and "strong" sustainability[5]. Where trade-offs between competing sustainability objectives are acceptable it is possible to pursue "weak sustainability". In such a situation, sustainability can be viewed as enhanced if the gain in one dimension more than offsets the loss in another. For instance, the wealth-reducing effect on future generations

5. *See OECD (2001),* op. cit., *p. 9, for a discussion of "weak"* vs. *"strong" sustainability, including some pointers to other literature.*

of higher gasoline prices due to dwindling oil reserves could be offset by increases in man-made capital that is bequeathed by the present generation, such as substantially more fuel-efficient cars. On the other hand, where the potential for substituting one goal for another is absent, the only option available is "strong" sustainability. An example is the environmental capital represented by the ozone layer. Sustainability in this case requires that the layer not be allowed to diminish below a threshold safe for humanity, and there is no known way to offset violation of that objective through investment in some other dimension of sustainability.

In general, where strong sustainability applies, minimum benchmarks need to be set. In the environmental dimension, for instance, it might be decided in a given country that a certain minimum capacity to absorb carbon dioxide emissions through forest resources must be maintained, even though the cost in lost output from forest-based industries is very high. In the social dimension, it might be decided that a certain percentage of the population must have access to electricity. In the economy, it might be decided that a certain degree of oil import dependence shall not be exceeded. In all these cases, society imposes constraints in the energy sector in order to maintain critical quantities of certain public goods associated with sustainability.

The principle of weak sustainability can be helpful in clarifying the nature of choices available in certain situations. Where it is acceptable, weak sustainability allows for arrangements that can advance sustainable development efficiently and consistently. Perhaps the best illustration is the "internalisation of externalities". Here it is assumed that the substitution of goals is acceptable and the government is allowed to set the terms on which the trade-off is made. If the cost to producers and consumers contributing to a given kind of pollution is increased through a pollution tax, less pollution will result. The polluters will install new technologies and reduce consumption of the now-more-expensive product causing the pollution. But the system still allows for some pollution, if consumers are willing to pay the price. The resulting tax funds can then be used to offset the impact of the pollution on long-run sustainability.

Properly applied, the principle of weak sustainability helps to establish consistency in policy-making through benefit-cost analysis. On the other hand, weak sustainability has its limitations. Benefit-cost analysis assumes that variables associated with all three of the sustainability dimensions can be expressed in a standard unit of account – usually money. This is often difficult in the environmental and social dimensions, as there is disagreement on how to value costs and benefits in these areas. Furthermore, while benefit-cost analysis is a powerful tool in situations in which trade-offs can be made on a marginal and gradual basis, it is often unconvincing in cases where fundamental choices and large changes have to be made or where irreversible damage will occur. Examples are decisions whether an animal species survives or whether a country with a vulnerable shoreline will continue to be habitable.

Those who argue in favour of using trade-offs assume that it is possible to know the probability distributions and damage functions that apply to sustainability risks and that such risks are reversible. Those who argue for the setting of physical thresholds find these assumptions questionable.

The way out of the debate over weak and strong sustainability seems obvious in general terms: both approaches should be used. Although drawing the line between them is often difficult. Which one is appropriate depends on the situation. For example, an action to support biodiversity might well specify how many of certain types of natural habitats are to be protected. On the other hand, dealing with sulphur dioxide emissions might best be done indirectly by influencing prices or establishing trading instruments.

Strategic Direction for Energy Policymaking

The response of energy policy-makers to the challenge of sustainable development is at a critical juncture. While the ideas mentioned above hold great potential, and some progress has been made, the adaptation of energy policy to a new framework will not be easy to achieve. A strategic sense of direction in regard to key goals is needed. The following criteria are proposed in order to provide a first orientation:

- Energy policies must balance the economic, social and environmental dimensions of sustainable development.
- They must contribute to the management of risk and the improvement of flexibility, in order to avoid serious disruptions of the energy system and the economic, social and environmental systems in which it functions.
- Energy policies should result from processes in which information and research are consciously managed and decision-making is well integrated, with broad stakeholder involvement.

ISSUES IN THE FORMULATION OF SUSTAINABLE DEVELOPMENT POLICY

Dimensions of Sustainability in the Energy Sector

Economic sustainability requires strong and durable economic growth with financial stability and a low and stable inflation. Environmental sustainability is centred on clean air and water, healthy soils and the resilience of biophysical systems. Social sustainability incorporates a range of values, such as equity, high employment, stability in social and cultural systems and participatory democracy.

These three dimensions of sustainable development, and the particular views that accompany each of them, have been developed, understood and integrated into policymaking to differing extents. The most highly developed is economic sustainability. Continuous growth in *per capita* real GDP has been widely agreed upon as the primary goal of economic policy[6]. The relevance of energy issues is obvious here. Continuous growth is achievable only if the security and affordability of energy supply are maintained. More difficult is the relation between energy and the other dimensions of sustainability. It is much harder to define a comprehensive policy in relation to environmental sustainability. Care must be used to define quantitative thresholds, the surpassing of which could pose threats to sustainable development. Examples are limitations on emissions of greenhouse gases or airborne pollutants with negative health impacts. The energy sector plays a key role in this area, *inter alia* because of the large fraction of greenhouse-gas emissions from fossil-fuel combustion. The social dimension is the most elusive of the three. Its relationship with the consumption and production of energy is discussed below.

6. The original charter of the OECD of 1961 emphasises its commitment to "sustained economic growth" in Member countries.

Box 1

Unsustainable Energy Use in History

The British economist William Stanley Jevons predicted the exhaustion of coal reserves as a major constraint on sustainable growth more than a hundred years ago. But the most serious sustainability challenges in the energy sector are not posed by the scarcity of non-renewable resources. They are the environmental and social side-effects of energy consumption or the lack of it. And this is not a new phenomenon. Unsustainable energy use looms large in human history. Examples are the near complete deforestation of the Hellenic peninsula largely completed in antiquity, and of the Spanish mesa and the British Isles in the 16th century. Since the beginning of the coal age, London fog – due to the burning of low-quality coal – has been infamous, with serious impacts on morbidity and mortality.

Claims that "200 years ago the energy system was sustainable" need to be treated with caution. It is instructive to compare the situation today, when 6 billion people aspire to Western living standards, with that at the time of the Industrial Revolution, when little more than one billion people populated the earth. Not only has population grown dramatically, but the magnitude and composition of annual per capita *energy consumption has increased from 0.9 MWh (0.88 MWh from coal) in 1860 to 19.1 MWh in 1998. In 1998 only 5.3 MWh were contributed by coal[7]. The main driver other than population growth has been, of course, world GDP, which grew from 470 in 1870 (indexed at 100 in the year 1500) to 11,664 in 1992. According to the same index* per capita *income grew from 117 in 1820 to 942 in 1992. In the Twentieth Century more energy was used than in all of preceding human history, or ten times as much as in the 1,000 years before 1900[8].*

7. *Joel E. Cohen (1995),* How Many People Can the Earth Support?, *p. 99.*

8. *J. R. McNeill (2000),* Something New Under the Sun: an Environmental History of the Twentieth-Century World, *p. 15.*

Sustainability issues with strong implications for the energy sector can arise in many different forms and are by no means new, as illustrated in the box above. The idea that human activities contain the seeds of forces that will eventually frustrate them has challenged thinkers over the centuries. What is new is that the question is being posed as a major policy paradigm on a global scale.

Economic Sustainability and Energy Supply Security

The importance of energy for economic development makes safe, universal and affordable access to it a necessary condition for sustainable development. Energy supply interacts with a broad set of development issues, in that it is intimately linked to certain modes of production and to lifestyles. The Industrial Revolution can be viewed as the history of a dramatic shift in energy sources. In a transition from an economy essentially based on agriculture to one based on industrial manufacture, production switched from dependency on human and animal muscle power, supplemented by wind, water and fire (from biomass and coal), to a massive dependency on fossil fuels. This huge shift from renewable energy sources to non-renewables precipitated today's sustainability concerns.

The long trend of rising fossil fuel use that began with the Industrial Revolution was thrown in question for the first time in the early 1970s. The two oil-shocks of the decade underlined the vulnerability of the energy sector in OECD countries and led to new initiatives in energy diversification, energy efficiency, energy conservation and emergency response measures. At the same time, widespread scepticism arose about future resource availability, especially oil, even though these fears appear at present to be largely unfounded[9].

9. A representative example of such Malthusian predictions of doom in this period was the report to the Club of Rome *by Dennis L. Meadows and others in 1972 with the programmatic title* The Limits of Growth. *In the meantime, faith in the power of technological and scientific progress to expand the frontiers of economic growth has remarkably increased. Undirected economic growth, rather than the lack of it, is now frequently seen as the principal challenge to sustainable development because it threatens environmental or social boundary conditions.*

The availability of oil reserves has been assessed with the help of an analytical tool called the "Hubbert curve," which indicates the rate at which reserves are transformed over time. The area under the Hubbert curve corresponds to total available reserves (see example below). Based on a reading of the curve alone, the IEA's *World Energy Outlook 1998* stated that oil production would peak as early as 2008/2009. When technological progress is accounted for, the area under the curve expands and, in the 1998 *Outlook*, changes its shape, with the right-hand tail becoming thicker[10].

Figure 3

Generalised Hubbert Curve

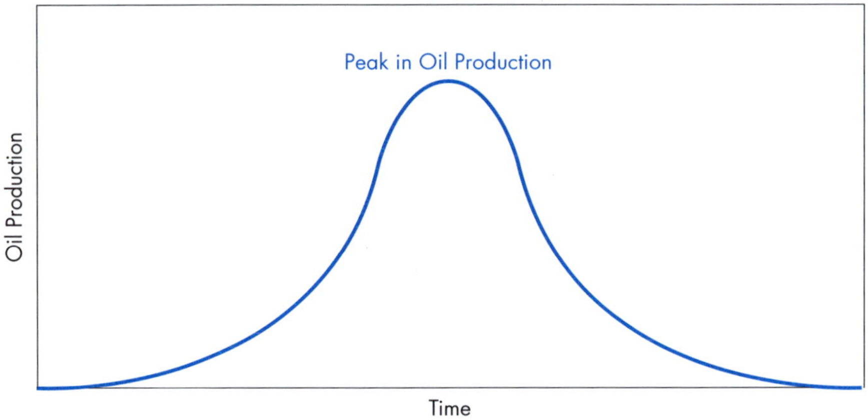

Note: IEA (1998), World Energy Outlook 1998, *p. 96.*

Taking technical progress into account leads to substantially more optimistic prospects. The *World Energy Assessment* assumes 45 years of physical availability (based on today's demand and supply) and 95 years in a dynamic perspective, where both would change[11]. The US Geological Service, which collaborated closely with the team preparing the IEA *World Energy Outlook 2000,*

---

10. *IEA (1998),* World Energy Outlook 1998, *p. 96f.*

11. *Presentation by Professor José Goldemberg at the International Energy Agency, Paris, 13 November 2000*

assumes a comparable figure, suggesting that available reserves could satisfy demand at the 1997 level for 124 years[12].

A declining share of oil in the value of total imports into IEA/OECD countries (see graph below) has also to some extent calmed the fears of threats to economic sustainability[13]. In IEA/OECD countries, oil's share of total imports in value terms has declined from 13 to 4 per cent between 1980 and 1997. However, caution is warranted in interpreting these numbers because they are measured in US dollars. During the same period exchange rates declined for all IEA/OECD countries (except, of course, the United States). This decline in the dollar exchange rate has massively contributed to the declining share of oil in the value of total imports. While structural changes have also taken place, their magnitude is more modest than the unadjusted trade figures would suggest. Recent oil price rises have also reversed the trend and increased once more oil's share in the value of total imports in OECD countries.

Energy diversification and emergency response mechanisms further contribute to stabilising supply security and thus economic sustainability. While threats to economic sustainability through energy supply disruptions still exist, they are better understood than in the 1970s and, at least to some extent, already taken into account in energy policy making (see page 51 and Chapter 3 for a discussion of supply risk and its management in the IEA context).

Environmental Sustainability and Greenhouse Gas Emissions

Environmental side-effects of energy consumption and their impacts on human health are as old as urban conglomerations. The soot-blackened walls of pre-historic human dwellings in caves suggest that they go back much further. The environmental historian J. R. McNeill reports that a quarter of all deaths in coal-powered Victorian England was due to lung diseases[14]. Some things

12. *IEA (2000),* World Energy Outlook 2000, *p. 46.*
13. *IEA (2000),* Oil Prices and Gasoline Taxes: Some Economic Facts, *EAD Working Paper, p. 2.*
14. *McNeill (2000),* op. cit., *p. 58.*

Figure 4

Share of Oil Imports in IEA Countries and the US$ Exchange Rate

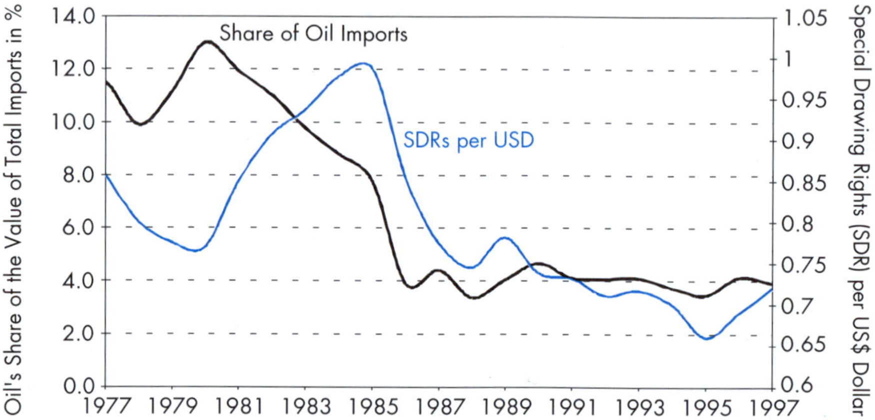

Note: *IEA (2000)*, Oil Prices and Gasoline Taxes: Some Economic Facts, *EAD Working Paper, p. 2.*

have improved since then. A gradual switch from coal to oil to gas has reduced environmental effects per unit of energy consumed and this amelioration is still ongoing.

But things are far from well. Despite progress here and there, on average local air pollution has increased substantially over the last century. London fog is estimated to have killed as many as 4,000 people during early December 1952, an event that defined a turning point in the city's energy policy, with a switch from coal to oil. Other modern examples of serious levels of energy-related air pollution are found in Pittsburgh, Athens, Los Angeles and Germany's Ruhr area. In the first half of the Twentieth Century, pollution in the form of particulates and sulphur dioxide was mainly related to coal consumption in heavy industry. In the second half of the century, ozone-producing volatile organic compounds from automotive traffic constituted the main public health problem from energy-related air pollution[15].

15. McNeill (2000), op. cit., *p. 87. McNeill also reports that air quality over Germany's Ruhr area, as well as the quality of its agricultural produce, improved markedly during France's occupation of the region in 1923, when all industrial activity was halted due to the ensuing general strike.*

Over the last several decades, heightened concerns about the environmental effects of energy use led to a first round of environmental regulations primarily aimed at local and regional airborne emissions. At present, movement in this area is driven by the efforts of developed countries and those in transition to reduce their greenhouse gas emissions, roughly four-fifths of which result from the combustion of fossil fuels. The graph below indicates the magnitude of this challenge. It shows the commitments of Annex B countries and the emissions projected in the Reference Scenario of the IEA *World Energy Outlook 2000*[16].

Dealing with the challenge of climate change involves many of the issues associated with sustainable development in general – a multitude of actors, a global dimension, interactions between the environmental, the economic and the social spheres, large threats of potential disruption and great uncertainty concerning the effects as well as costs of prevention. The Kyoto Protocol on greenhouse gas reductions and the Montreal Protocol on the phase-out of ozone-destroying substances are among the first global efforts in which countries have committed themselves to take costly actions in order to achieve global benefits. Climate change is certainly not the only sustainability challenge. However, its magnitude and complexity make it a case at the centre of the paradigm. The success or failure of the Kyoto Protocol will constitute something of an indicator of the commitment of the global community to sustainable development.

At the same time, it is important to acknowledge that the environmental effects of energy production and consumption are not confined to airborne pollutants or climate-changing greenhouse gas emissions. The extraction of fossil fuels can scar local environments. Oil transport by tankers has caused some of the worst environmental catastrophes in recent years, although

16. Annex B countries under the Kyoto Protocol have committed themselves to reduce their climate-relevant greenhouse gas emissions by 2010 to 6 per cent below their 1990 levels on average. All IEA Member countries except Turkey are Members of Annex B.

Figure 5

CO_2 Emissions for Annex B Countries

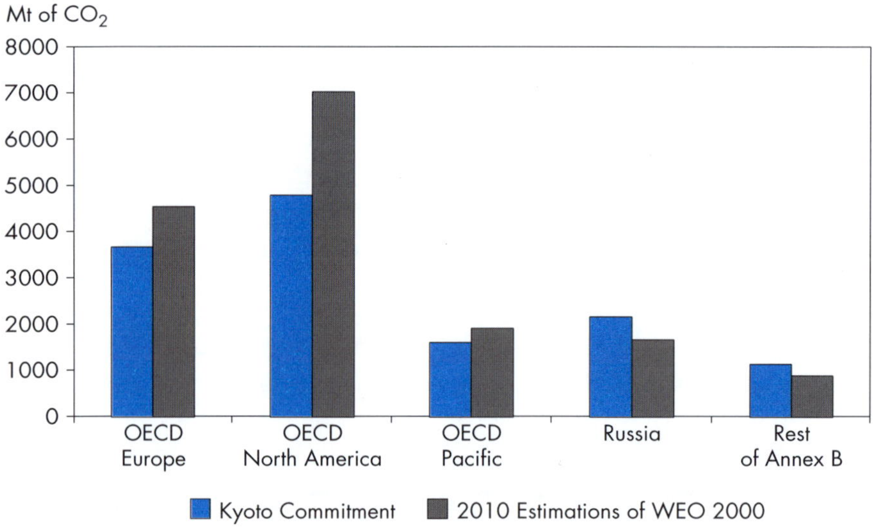

Note: Based on IEA (2000), op. cit., *Part D: Tables for Reference Scenario Projections, p. 349-418.*

accidental oil spills contribute less than 10 per cent of the total oil ending up in the oceans – most of it non-accidental[17].

There are also problems with non-fossil fuels. Dams cause salinisation, withhold silt and can cause large amounts of greenhouse gas emissions due to decomposing biomass and the drainage of wetlands. Nuclear energy poses potential threats due to the long-term radiation of nuclear waste and possible accidents. Even renewables can have negative environmental effects on land-use and through noise (wind), de-forestation (biomass), high toxicity (photovoltaics manufacture) or the threat of accidents (hydrogen). Any form of energy production and consumption that occurs on a sufficiently large scale is likely to be at odds eventually with environmental sustainability given the finiteness of the natural biosphere and the exponential growth of human activities[18].

17. Ocean Planet: Oil Pollution at http://www1.imo.org/oilweb/default.htm#un tech.

Social Sustainability and the Energy Sector

The social dimension can be approached both in terms of wealth distribution and of social organisation. The now dominant approach is to view social sustainability as the maintenance and enhancement of social capital. A forthcoming OECD publication defines it as "networks together with shared norms, values and understandings which facilitate co-operation within or among groups."[19] Social capital is thus the sum of the implicit rules, traditions and values, the knowledge and the relationships that are constructed through historical time and help a society to run smoothly day-in and day-out. Social capital lowers transaction costs, increases creativity and innovation, and improves the welfare of individuals and communities.

A priori the energy sector does not appear to have a special role in this dimension. However, when some energy-related activities are considered in the context of the history and social development of certain regions – notably coal mining in Germany or the United Kingdom – a case can be made that the link between energy and social capital needs to be considered.

Social models and hence the nature of social capital differ vastly across countries and population groups. The contribution of energy to social capital will be quite different in, say, Europe, Asia and Africa. Given the absence of more formal social infrastructures, that contribution is likely to be more important in developing countries than in OECD countries. Improvements in the provision of energy, in particular the availability of electricity, have demonstrable effects on the quality of life. Electricity

18. J. R. McNeill emphatically claims that "Oil... was the single most important factor in shaping environmental history after the 1950s..." only to continue "Which is not to say other energy regimes would have been environmentally neutral. One based on muscle and biomass (the Haitian model) would have stripped the world of combustible vegetation. One based on coal (the Polish model) would have markedly increased air pollution. One based on nuclear energy ... would have run greater risks of meltdowns and committed the world to millennia of management of more lethal wastes. As for one based on photovoltaics, wind power, and fuel cells – we don't know (yet) what that might entail." Op. cit., *p. 305-6. For a detailed discussion of environmental costs and benefits of renewable energy see IEA (1999),* Benign Energy? The Environmental Implications of Renewables.

19. OECD (2001 forthcoming), Reconciling New Economies and Societies: The Role of Human and Social Capital.

improves possibilities for education, for public health, for mobility and for economic productivity.

The close link between the social dimension and energy use in developing countries is displayed by Figure 6 below. The figure shows the link between the energy mix and social parameters such as location and *per capita* income. At the same time, the difference in the energy mix has a large incidence on issues such as health. Rich urban households cover roughly 80 per cent of their energy needs with clean-burning gas, while poor rural households use up to 90 per cent firewood that causes significant outdoor and indoor pollution.

Such considerations were important in OECD countries in the past, but nearly universal access to energy systems has made them recede into the background and they are no longer priority concerns for policymakers today[20]. Thus further consideration of them in this book focuses on those aspects relevant to developing countries (see Chapter 9).

Three Approaches to Balancing the Sustainability Dimensions in the Energy Sector

The problem of balancing the three dimensions of sustainable development is explored in this section by considering it in relation to three policy options:

- the maintenance of per capita incomes through time,
- the internalisation of energy externalities, and
- the removal of energy subsidies.

Maintaining *per capita* Incomes

"However it may be defined in detail, achieving sustainable development necessarily entails creating and maintaining wealth..."

20. There are exceptions. The dramatic reactions of European truck drivers to the sudden increase in fuel costs in the summer of 2000 indicate that social conflict over energy in Europe has not disappeared.

Figure 6

Energy Use in India by Social Group

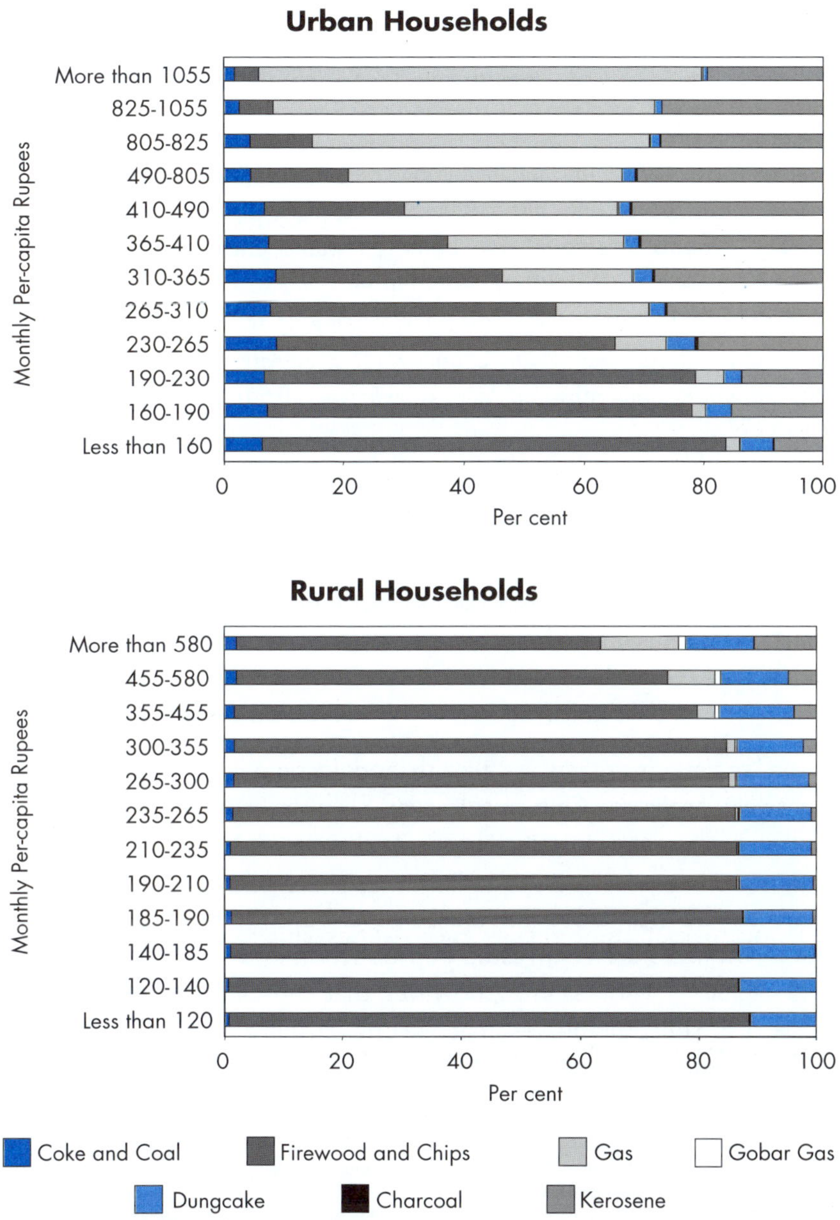

Note: IEA (2000), World Energy Outlook 2000, *p. 318.*

assert the authors of a World Bank report on the measurement of national wealth[21]. Maintaining standards of living and welfare is the essence of sustainability.

The income on which people live and the goods and services they consume flow from wealth produced from a combination of economic, environmental and social assets.

Man-made assets include buildings, machinery and infrastructure in the form of roads, ports and airports, water supplies, pipelines, electrical networks. Natural assets include both renewable and non-renewable resources, as well as the environment. Human and social assets include education, health, knowledge and understanding of science, technology, culture and human behaviour, capacity for creativity and innovation, ability to store and communicate knowledge, institutions and social networks[22].

Man-made assets and natural resources that are traded on markets can be valued in monetary units and kept track of in terms of Gross Domestic Product accounting. The values of these assets and their services can therefore be easily compared, though there are of course difficulties in obtaining dependable estimates in some situations. The evaluation of renewable environmental resources and of social assets is relatively more difficult (see below the section on externalities).

How a society maintains its wealth and income levels has been extensively analysed. One result that is often cited is another illustration of weak sustainability – known as "Hartwick's rule". Hartwick's theoretical analysis indicates that a constant consumption pattern can be maintained if resource rents from the exploitation of natural resources are re-invested in production in the form of man-made capital[23]. This assumes, however, that the different forms of capital are substitutable – that strong sustainability rules are unnecessary for the maintenance of those environmental and social

21. *Kirk Hamilton and Michael Clemens (1998),* Genuine Savings Rates in Developing Countries, *World Bank, Washington, DC; http://www-esd.worldbank.org/eei; p. 1.*

22. *Nuclear Energy Agency (2000),* Nuclear energy in a sustainable development perspective, *p. 11.*

23. *OECD (2001),* op. cit., *p. 9.*

assets. In reality many of these assets are non-substitutable. Hartwick's rule is thus useful only for a limited number of applications.

Nevertheless the principle of non-declining wealth is a convenient minimum criterion which can be applied to issues where trade-offs are acceptable. An important example with particular relevance to the energy sector regards the trade-off between marketable natural resources embodied in oil reserves and financial wealth. In this case, the objective of achieving the highest possible *per capita* income over time in the nation that owns the oil reserves raises important issues.

It has been dealt with in Norway, where the principle of maintaining wealth for future generations has determined the use of the revenues from the country's large oil and gas reserves[24]. Norwegian government revenues from oil and gas accounted for 8 per cent of GDP in 1997 and for 4 per cent in 1998. Earlier in the 1990s Norwegian petroleum assets were estimated to be worth between $^1/_2$ and $2\,^1/_2$ times one year's GDP. Generational-accounting is common practice in Norway and rules on government spending have been established to distribute shares of oil revenues to future generations. This includes a rule on government budgeting. The fiscal deficit, net of petroleum revenues is not allowed to exceed the estimated return on the remaining petroleum wealth[25].

The principal means of transmitting current wealth to future generations is the Government Petroleum Fund, established in 1990. The Fund absorbs the government's fiscal surplus, which has been on the order of 5 per cent of GDP per year since 1996. In 1998 the Fund contained NKr 136 billion (US$ 14.4 billion) or 14 per cent of GDP. It is currently managed by the Norwegian Central Bank. The bank invests it only in assets denominated in foreign currency in

24. *Paul van den Noord and Ann Vourc'h (1999),* Sustainable Economic Growth: Natural Resources and the Environment in Norway.

25. Ibid., *p. 11f.*

order to offset exchange rate movements on oil export revenue and to offset the risk of a combined fall in the oil price and the value of associated domestic assets. Future generations in Norway will also benefit through the legacy of public assets that have been built with oil revenues. Oil revenues benefit the present generation through lower taxes and a generous welfare system.

The Norwegian example illustrates the role that can be played by the principle of non-declining wealth over the long run. Those who favour free-market solutions would, of course, argue that distributing the whole of the petroleum wealth to the private sector would create as much or more wealth for future generations. Private markets for capital are, after all, considered the most efficient means to allocate scarce funds to generate the highest return. Private investors would create wealth by investing their surplus funds in projects with the highest returns, while diversifying financial risk in the process.

There are counter-arguments to this position. One is that only a national co-ordinating body, such as a central bank, can deal effectively with system-wide risk through diversification. In addition, there is no guarantee that private savings rates would be as high as public savings rates. It is entirely possible that the current generation could go on an unsustainable spending binge, creating excess demand and inflation in one period and a hangover of overcapacity and recession in the next. This was the experience in the Netherlands after the initial exploitation of its gas reserves – the experience has been labelled as the "Dutch disease". While maintaining *per capita* wealth, or gradually increasing it through time, may seem like a dull idea it can be quite important in practice.

The Internalisation of Energy Externalities

Measurement and Conceptual Problems

Concern about damages to material structures, human morbidity and mortality from air-borne emissions due to energy consumption has led to an unprecedented effort to quantify these damages – a concept known as the "external costs of energy". It has been argued that this calculation would allow the costs of harmful emissions to be taken into account in decisions concerning the choice of energy systems and fuels. By assessing externalities over the life-cycle of different energy sources and "internalising" them into economic decisions, it would be possible to have an energy system at the lowest possible full cost – "full" in the sense that all environmental and social costs would be accounted for. In practice, the means to achieve this would be a "Pigovian tax" (named after its first proponent, the British economist A. C. Pigou). Internalising energy externalities would thus efficiently integrate the environmental and economic dimensions and allow an important step toward sustainable development.

In the 1990s great progress was made in the quantitative measurement of energy externalities from electricity production through the combined efforts of the European Commission, the US Department of Energy and others. At the same time, analysts are very aware of the difficulties in obtaining comprehensive and accurate measurements of external effects and especially with using them in an international context. Measured values are in many instances highly dependent on local and national determinants, such as population density or climate, and on specific assumptions subject to the judgement of the researchers involved. Furthermore, it is difficult to capture the effect of energy sources with very high potential damages but low or uncertain probabilities, such as nuclear accidents or climate change.

The table below compares the results of three large studies of energy externalities[26]. These studies demonstrate the feasibility of externality measurement in principle, but they also show large differences in the evaluation of damage costs. The differences occur because of uncertainty over the monetary valuation of the damages that underlie the aggregate estimates, as well as the

26. *Keppler and Kram (1996),* Energy Markets — Full Cost Pricing. *Based on European Commission (1995),* ExternE: Externalities of Energy; *Hagler-Bailly (1995),* New York Environmental Externalities Cost Study; *Nick Eyre (1994),* Personal Communication; *Oak Ridge National Laboratory (ORNL) – Resources for the Future (RFF) (1994),* Fuel Cycle Externalities.

Table 1

Damage Cost Estimates from Three Recent Studies (US$ per tonne)

Study		SO_2	NO_x	Particulates
ExternE	High estimate	6,050	12,610	16,060
	Low estimate	4,140	0	16,060
ORNL/RFF Study Southeast Reference Site	High estimate	1,002	2,003	4,004
	Mid estimate	60	120	1,900
	Low estimate	10	90	850
New York State Externalities Study Central Estimates	Urban	1,200	1,100	43,800
	Suburban	800	900	7,700
	Rural	700	900	3,200

underlying biophysical relationships associated with the polluting substances. They draw attention to the limits of the approach, which is most easily applicable to effects that are clearly determinable and measurable.

In order to quantify and monetise preferences for the public goods that are damaged through energy use, and thus derive damage values for environmental externalities, two conditions need to be fulfilled. The relations among production, dose of emissions and environmental impact must be identifiable and quantifiable. Consumer tastes for public goods – most importantly environmental quality – must be both stable and comparable with tastes for marketed commodities. There are situations in which sufficiently precise causal relations can be established and where the effects on wellbeing can be satisfactorily expressed in monetary terms. This is true mainly for local and regional externalities with clearly identifiable effects on human health and materials. In other situations, these conditions do not apply. In the case of climate change, for instance, neither condition is fulfilled in any comprehensive way and there exists little basis to measure and monetise relevant costs.

Environmental taxes raise other issues. One argument against Pigovian taxes points to their distributional impacts. The imposition of a Pigovian tax assumes that the property rights of environmental resources such as clean air belong to the general public and not to the polluter. This seems a reasonable assumption, but a long history of allowing polluters to infringe on those property rights has affected public attitudes and often created customary law in favour of the polluter. Changing this situation will affect income distribution (though it leads to an increase in total welfare) and this can provoke massive resistance. Opponents of change point to concerns about international competitiveness and the possible relocation of polluting industries to so-called "pollution havens", countries with less stringent environmental policies.

Another important issue is how well the Pigovian approach serves the objectives of sustainable development. One limit of this approach is that it is static. It is based on the assumption of a set of given relationships, such as damage functions and demand- and supply-curves, an assumption that leads to the measurement problems already discussed. More importantly, static optimisation focussed on finding "optimal" prices and quantities – is not necessarily the right goal in the fast-changing context of sustainable development in the energy sector, especially in light of the rapid market restructuring that is now underway. A dynamic environment of this sort requires a broader view, and additional policy instruments that can deal with the structural parameters of markets such as competition, consumer awareness and new technologies[27].

A dynamic approach to energy externalities would focus on the management of change. It would proceed from the view that externalities are often new phenomena to which societies must adapt, and that internalisation can take place in many ways. The dynamic and static approaches are of course linked. A change in

27. *See Keppler (1998),* Externalities, Fixed Costs and Information *on the methodological basis of the theory of externalities and the rationales for the dynamic approach.*

structural parameters will also change prices and quantities over time and a dynamic approach does not necessarily disavow the Pigovian policy paradigm. Prices in the form of environmental taxes for the consumption of public goods and the prescription of certain technologies or quantities through regulation have their roles to play. But the focus needs to change and the paradigm of static optimisation needs to be inserted into the larger framework of managing change.

Another way of understanding a dynamic approach to externalities is to relate it to the different dimensions of sustainable development. An energy externality is one effect of an economic activity on the environment. A policy instrument – generally a fiscal instrument – is used to integrate (or internalise) it back into the economic dimension. Internalising the negative impacts of one dimension of sustainable development into another can be as simple as establishing channels of communication between those responsible for the externality and those affected by it.

The nature of the internalisation mechanism should correspond to the structure of the problem. Information about externalities is frequently vague and ill understood, due to their newness. For this reason, research and development policies are important, as are new approaches to preference formation. R&D is useful because it leads to new options with reduced negative impact on sustainability. Preference formation refers to all those societal processes in which values and perceptions are sharpened and made more precise. This includes public discussion, the press and the political processes. Preference formation occurs when a "gut feeling", an "intuition", an "unease", a "fear" or a "preoccupation" is translated into more precise expressions of welfare loss that can subsequently be linked to specific instances.

Under the dynamic approach to sustainability, solutions to problems are inserted into the processes of development in the energy sector. They not only respond to change but are in themselves part of it by influencing such structural characteristics as technological parameters, the degree of competition, the form of regulatory oversight, product characteristics, infrastructure provision and pricing and consumer preferences.

Trade-offs at the margin between different energy sources – which is the general result of the Pigovian approach – are important. They do not, however, address the large-scale systemic impacts that policies for sustainable development try to address. The World Bank expresses something similar when it points to the need to proceed from the internalisation of externalities to "considering the environment an integral part of development[28]."

Progress in Reducing Pollution

IEA/OECD countries have made considerable progress in recent years in reducing the emissions of local and regional pollutants, such as particulates, sulphur-dioxide or nitrogen oxide (see graph below). However, this progress has been made mainly through regulatory efforts, such as the Large Power Plant Directive in the European Union or the Clean Air Act in the United States, rather than through fiscal internalisation.

Unfortunately the same progress has not yet been made in developing countries, where the effects of energy-related emissions remain dramatic. In fact, the poor suffer disproportionately from health damages due to the use of coal, dung and biomass in open furnaces for cooking and heat. The *World Energy Assessment* estimates that poor indoor air quality due to burning solid fuels is responsible for about 2 million premature deaths a year, mainly among women and children[29].

A case in point is China, which is heavily coal-dependent and where sulphur dioxide and particulates heavily damage human health, agriculture and structures. One study estimates the total costs at US$ 13 billion per year[30]. Sulphur dioxide (SO_2) emissions from

---

28. *World Bank (2000),* Toward an Environment Strategy for the World Bank Group, *p. 4.*

29. *UNDP, UNDESA and WEC (2000),* op. cit., *p. 15.*

30. *Battelle Memorial Institute (1998),* China's Electric Power Options: An Analysis of Economic and Environmental Costs.

Figure 7

Progress in Reducing Emissions from Power Generation in OECD Countries (Index: 1990 = 100)

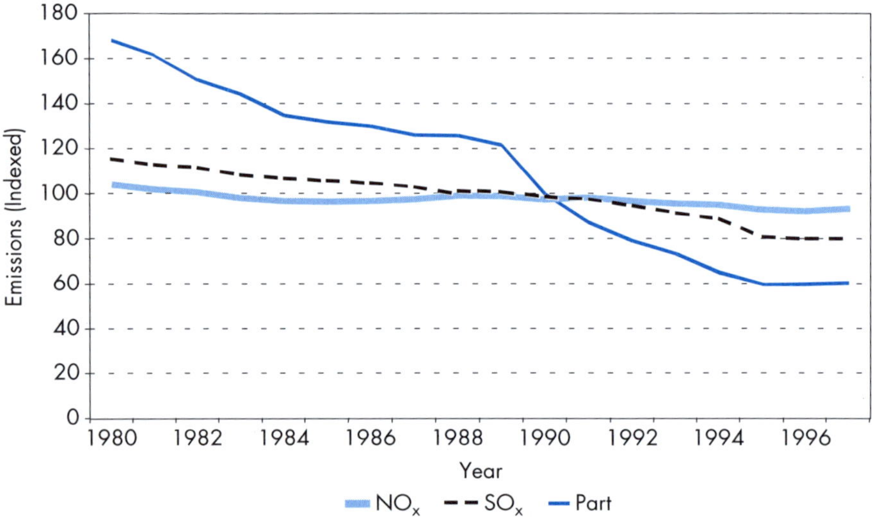

Note: Compilation on the basis of OECD (1999), OECD Environmental Data – Compendium 1999, *Paris.*

Chinese industries and power plants affect 30 per cent of China's territory in the form of acid rain, which reaches as far as Korea and Japan. While the Chinese national average is about two tonnes of SO_2 per square kilometre, comparable to that in the United States, the Chongquing urban area is reported to have a staggering 600 tonnes per square kilometre, with effects on the environment and public health to match.

Harmful environmental impacts in developing countries are widespread. The table below shows emissions levels in a number of large cities that in most cases exceed the World Health Organisation air-quality guideline levels.

In these countries improving energy efficiency, switching fuels from biomass, dung and coal to gas and electricity and basic emissions controls could achieve substantial reductions at limited costs. The transfer of more efficient and less polluting energy technologies from OECD countries can play an important role in this context. Under the right circumstances, such transfers could deliver benefits in the economic as well as in the environmental dimension of sustainable development (Chapter 9 reports in detail on the transfer of efficient energy technologies to developing countries).

Inevitably, efforts to establish environmentally sustainable energy sectors in the developing world will be complicated by those countries' aspirations to consumption levels like those in OECD countries, aspirations that can hardly be denied from a moral point of view. At the same time, it is hard to imagine the effect – and the magnitude of the energy externalities – that would come with European or American levels of automobile ownership and other consumer goods. In 1997, there were three cars per 1,000 inhabitants in China and five per 1,000 in India.

Table 2

Airborne Emissions of Major Pollutants in Developing Countries (1995)

Country	City	City Population (1,000s)	Total Suspended Particulates (mg per m^2)	Sulphur Dioxide (mg per m^3)	Nitrogen Dioxide (micrograms per m^3)
China	Beijing	11,299	377	90	122
Russia	Moscow	9,269	100	109	N/A
India	Delhi	9,948	415	24	41
Indonesia	Jakarta	8,621	271	N/A	N/A
Iran	Tehran	6,836	248	209	N/A
South Africa	Capetown	2,671	N/A	21	72
Venezuela	Caracas	3,007	53	33	57
WHO Guideline			90	50	50

Source: IEA Compilation (2000).

The Removal of Perverse Energy Subsidies

Energy subsidies, particularly those that keep prices below cost and therefore encourage energy consumption, can impose a heavy burden on economic efficiency, environmental performance and government budgets. This burden has to be weighed against potential social benefits, such as helping to assure that the poor have adequate energy services, and a variety of environmental benefits. Subsidies for commercial fuels such as coal, kerosene or natural gas can be beneficial when they lead to a reduction in "predatory" biomass consumption – the clear-cutting of forests for the purpose of collecting firewood without re-forestation. Clear-cutting is totally unsustainable, because it degrades the resource base it relies on and contributes to the destruction of sinks for greenhouse gas emissions.

Economists have argued that the use of generalised energy subsidies to achieve social and environmental objectives often cost more than directly targeted measures. Where subsidies for energy consumption already exist, the gain in economic efficiency that would result from their removal could, for instance, be spent on more effective social programmes. The reduction in energy consumption associated with prices close to true costs brings environmental benefits through reduced emissions. Assuming the energy being subsidised is non-renewable, it reduces the rate of depletion of the stock of natural capital. Economic growth would be boosted through improved efficiency and reduction in the size of the government budget.

Direct subsidisation of the consumption of energy occurs mainly in developing countries, where the objective of alleviating poverty looms large. The subsidisation of energy may be justified in such situations because energy is a necessity; throughout the world, poor households pay a larger fraction of their incomes for energy than do the rich[31]. However, this argument does not always apply – in many developing countries, the poor have no access to

31. *UNDP, UNDESA and WEC (2000)*, op. cit., *p. 12.*

commercial energy even at subsidised prices. In such cases, energy subsidies are actually *regressive*. Energy subsidies benefit the poor only in those cases where they target the fuels that poor people actually use.

In a few cases, social security systems are in such a dismal state that energy subsidies, together with subsidies for food, are essentially the only way of providing support for low-income groups. Obviously in such cases, alternative support mechanisms should be established if energy subsidies are eliminated. As a general rule, where subsidies are used it is often better to provide them for access to energy services, by paying the costs of connection to an electricity system, rather than for consumption. The social benefits from energy consumption – education, information, public health, mobility – tend to be connected with a few, high-value uses, rather than with high energy consumption *per se*.

Even where subsidy removal is justified by a comprehensive benefit-cost analysis, resistance to removing energy subsidies can be very strong. This became clear from the discussion at several recent workshops on energy subsidies organised by the IEA and UNEP. In some cases the resistance comes from well-organised pressure groups, which include people and firms who would lose substantial benefits as a result of subsidy removal. Large energy consumers benefit disproportionately from energy subsidies[32].

An IEA study has been done to measure the costs of subsidies for energy consumption in eight large non-Member countries: China, India, Indonesia, Iran, Kazakhstan, Russia, South Africa and Venezuela[33]. On average, consumer prices in these countries are 20 per cent below their opportunity-cost or market-based reference levels, despite substantial progress in recent years to move toward more rational and market-based pricing. Subsidy removal in all eight countries would cut global energy consumption

32. *See IEA/UNEP (2001),* Energy Subsidy Reform and Sustainable Development: Challenges for Policymakers, Draft Synthesis Report, *p. 12-14.*

33. *IEA (1999),* Getting the Prices Right: Looking at Energy Subsidies.

by 3.5 per cent and global CO_2 emissions by 4.6 per cent. The study concentrated on the costs of subsidies, without quantifying potential losses from subsidy removal. The results are presented in the table below.

OECD countries tend to subsidise energy production mainly to protect domestic industries. Estimates for total energy subsidies in OECD countries hover around US$ 20 billion[34]. One third of this amount goes into price support for domestic coal, especially in Germany, in order to maintain competitiveness against low-cost importers. Other subsidies support the development and adoption of more energy-efficient technologies and renewable energies. The impact of those subsidies on environmental sustainability can thus be positive or negative. Their effect on economic sustainability is

Table 3

The Results of Subsidy Removal in Eight Developing Countries

	Average Subsidy (% of cost)	**Annual Economic Efficiency Gains** (% of GDP)	**% Reduction in Energy Consumption**	**% Reduction in CO_2 Emissions**
China	10.89	0.37	9.41	13.44
Russia	32.52	1.54	18.03	17.10
India	14.17	0.34	7.18	14.15
Indonesia	27.51	0.24	7.09	10.97
Iran	80.42	2.22	47.54	49.45
South Africa	6.41	0.10	6.35	8.11
Venezuela	57.57	1.17	24.94	26.07
Kazakhstan	18.23	0.98	19.22	22.76
Total Sample	21.12	0.73	12.80	15.96
World	N.a.	N.a.	3.50	4.59

34. *IEA (2000),* Energy Subsidies in OECD Countries, *p. 16.*

unequivocally negative. On the other hand, claims that energy subsidies contribute to social sustainability, through the maintenance of the social fabric and regional cohesion do have some substance. However, in view of the benefits provided by other deserving activities, the advantages of energy subsidies seem insufficient to deserve such lavish support.

Subsidy removal can also entail a number of other gains. Energy industries become more dynamic through improved transparency and accountability, through accelerated development of technology and through a more entrepreneurial approach to exploration, production, distribution and supply. Opening energy markets to competition and to foreign direct investment is incompatible with the maintenance of energy subsidies that maintain the *status quo*. Both OECD and most non-OECD countries, including all those referred to above, now pursue policies aimed at increasing the role of the market in the economy generally and in the provision of energy supplies. In most cases, subsidy reduction and the removal of price controls are central features of these policies (see also Chapters 4 on market reform and 9 on non-Member countries).

Risk Management

Risk Management and Sustainable Development

If sustainable development is to be "development that lasts", it must take account of the risks of breakdown. Large-scale risk in the energy sector can lead to irreversible effects that would constrain the economic, social and environmental options of future generations. To avoid or limit such risk, the energy system must be robust and resilient in response to shocks and errors. In the face of an inherently uncertain future, this calls for substantial investment in flexibility and in systems capable of responding to unexpected shocks. There is therefore a need to manage risk, a policy position that is well supported in light of the widely-held human desire to avoid risk and the high costs of sudden and discontinuous change.

Risk management is particularly important in the energy sector due to two factors that limit short-term options in response to change. First, energy services such as heating, transportation, lighting and the maintenance of food supplies are vital necessities that need to be maintained. Second, the planning, construction, management and decommissioning of energy installations and infrastructure stretch over long timeframes. The slow turnover of energy-specific capital stock creates large irreversibilities once a decision has been made; changing the way of doing things before existing structures are fully amortised involves huge costs.

Figure 8

Average Lifetimes for Energy-related Capital Stock[35]

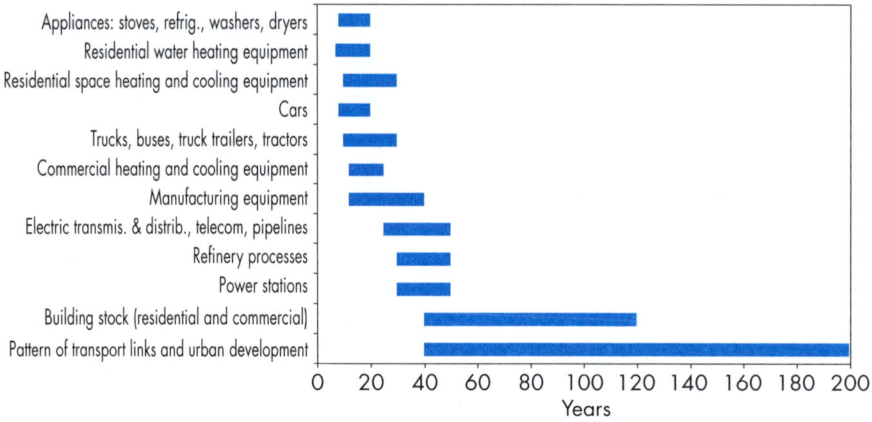

Note: IEA (2000), World Energy Outlook 2000, *p. 43.*

Inertia and rigidities due to these factors limit the flexibility of energy systems and expose them to disruption by unexpected shocks. Risk management is hardly new to the energy sector. There is a long history of private and public measures to increase the robustness and resilience of energy systems through insurance and other approaches that reduce the incidence of risk. Financial and technical risk, risks involved in exploration and resource

35. *IEA (2000),* World Energy Outlook 2000, *p. 43.*

depletion, the risks of environmental degradation and social disruption and, perhaps most significantly, risks to energy supply, are familiar features of the energy sector.

Risk Management in the Energy Sector

Multi-dimensional risk management, with a future-oriented long-term view, has been an important issue for some time. The classic example is the management of risks to energy supply, in particular the risk of interruptions in oil supply, in order to avoid economic and social disruptions.

Supply risk management has short-term and long-term components. In the immediate short-term, a flexible response system can tap reserves to soothe financial speculation and over-reaction in sensitive markets. International co-operation provides extra room for manoeuvre. Short-term responsiveness is important due to the extreme sensitivity of economies to perceived volatility. Sudden price changes can be more harmful than high prices *per se*.

In the medium and long term, diversification is the key measure to reducing risks and to maintaining economic activity. This means the diversification of both oil suppliers and energy sources. "Diversity, efficiency and flexibility within the energy sector..." are part of the *Shared Goals* of IEA Member countries. Technical flexibility, such as dual-firing capacity or the availability of measures for restraining demand, also contributes to the robustness of the energy system (Chapter 3 provides a detailed discussion of this kind of risk management in the context of the IEA's emergency response system).

The commitments of Annex B countries to reduce greenhouse gas emissions under the Kyoto Protocol can be viewed as another example of risk management in response to a large-scale systemic risk that threatens future generations. To help in achieving these commitments, the Kyoto Protocol created three "flexibility mechanisms", which increase the options for fulfilling the abatement commitments and reduce the total cost of compliance (see Chapter 8).

Figure 9

The Distribution of Global Oil Reserves

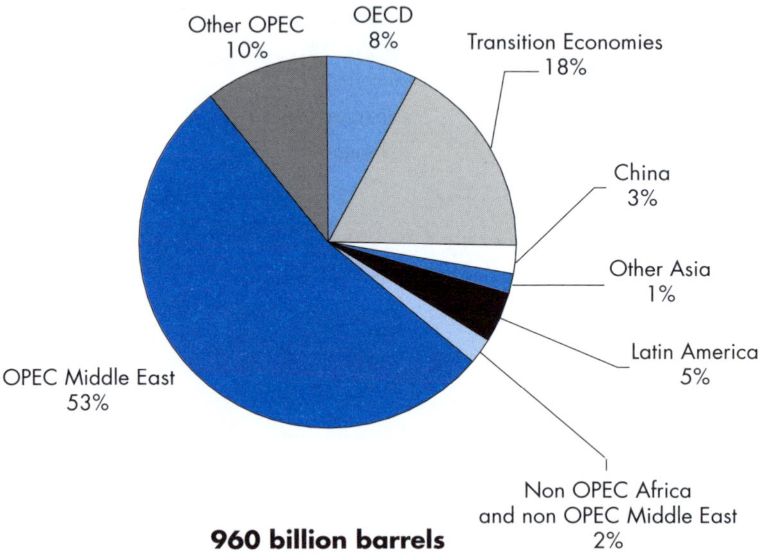

Source: IEA (2000), World Energy Outlook 2000, *p. 76.*

The classic form of risk management in the energy sector involves efforts to ensure the safe and accident-free operation of machinery and equipment for the extraction, transport and transformation of energy. Nuclear energy plays a special role in this context, both due to the catastrophic size of a potential accident, as well as due to the effort to actively assess and manage multi-dimensional risk in the nuclear industry. However, operating an LNG terminal, transporting petroleum or running a coal mine also requires the conscious assessment of system-wide resilience in response to human or technical failure in order to minimise risks for human safety and the environment.

The options for the private management of risk by producers and consumers thus play an important role in the energy sector. The purchase of different kinds of insurance and the use of futures markets are two major examples of private approaches to risk management.

Insurance and the Reversibility of Risk

Insuring a house against fire is relatively easy, as large amounts of data about the historic probabilities of fire hazards to comparable structures exist. Insuring against a unique, complex and irreversible event, such as climate change, depends on informed guesses about probabilities and impacts. It is very difficult to balance the cost of mitigating this kind of worldwide risk and the cost of doing so. The question of 'How much insurance?' must always be faced. Humans are risk averse, but they are also averse to paying for insurance without clearly perceptible benefits. This can create a serious problem in relation to sustainable development because the expected benefits of insurance decisions are much less clear to those who have to pay the present costs than is the case in classic personal or business insurance decisions. It is hard to sell the advantages of taking costly action on risks such as climate change, but, on certain issues in the sustainable development debate, there is also pressure for unrealistic amounts of insurance. Pursuing 'zero-risk' in relation to many objectives would imply impossibly high costs and thus endanger the economic dimension of sustainability[36].

A key issue in regard to how the insurance objective is approached is the question of reversibility. Privately insurable risks require "substitutability", for example between a house and a corresponding sum of money. The damage can be 'reversed' or compensated for in some way and this makes possible the use of a classic insurance instrument. On the other hand, it is unrealistic to expect that classic insurance instruments can be used to deal with large-scale, unique and irreversible risks. It is necessary to find ways to insure indirectly, by maintaining minimum levels of protection for certain important aspects of economic, social and environmental systems. The difference between insurable and uninsurable risks is parallel to the distinction between weak and strong sustainability discussed earlier in this chapter.

---

36. *In complex systems such as energy systems or civil aviation, the singular 'freak accident' can always strike in unforeseen ways. The costs of risk avoidance, however, rise over-proportionately as ever-smaller improvements toward this irreducible amount of risk are made. See René Amalberti (2000),* Dans les entreprises le risque zéro n'existe pas *and René Amalberti (1999),* Les effets pervers de l'ultra-sécurité.

Thus, while private insurance sold through markets is an important sustainability tool, there are obviously limits on its use. Markets are efficient in pooling comparable individual risks and in diversifying among mutually offsetting risks for private participants. Well-known examples are hedging in oil futures and portfolio diversification in energy investments[37]. In order to perform these functions, markets need explicit information upon which prices can be formed. Complex singular risks, such as a changing climate, a large nuclear accident or drastic action by OPEC, which figure large in concerns about sustainable development, defy explicit pricing. Government involvement is necessary and policy actions need to be decided upon in appropriate institutions that integrate relevant stakeholders. Both of these themes – the complexity of sustainable development in the energy sector and the role of institutions – are explored in the next section.

There is an intrinsic uncertainty about the impacts of technology, structural change and policy decisions that calls for caution, the hedging of bets and investment in back-up systems[38]. The management of large-scale risks cannot be neatly summed up in a number of easily implementable directives for action. No one has better expressed the dilemma posed by decision-making under circumstances of incomplete information and irreversible risk than Machiavelli in *The Prince*:

And in this it happens like physicians say of consumption, which when it begins to do its evil is easy to heal but difficult to recognise, but which in time – having neither been recognised in the beginning nor treated – becomes easy to recognise and difficult to heal. Like this it happens also in state

---

37. Price risk in particular can be hedged in oil futures markets. Making the appropriate tools available to companies for whom fuel costs are a large fraction of their expenses, such as transport businesses, would make these companies economically more sustainable.
Another practical example of risk management is the formation of a consortium of producers and consumers of photovoltaics equipment, who agree on minimum volumes of production at pre-negotiated prices in order to share risk and allow the exploitation of economies of scale and learning investments.

38. A good example of the need for caution is found in the history of urban development. Suburbs were seen by social planners in the first half of the Twentieth Century as a formidable means to switch to healthier and more comfortable life-styles. Issues of urban sprawl, local air pollution and gridlock, let alone oil supply security or climate change, had not even entered the discussion. See Edward Tenner (1997), Why Things Bite Back, *p. 335-338. As the title suggests the book provides a long list of examples of technological progress with unintended consequences, both positive and negative.*

Box 2

A Sustainability Lesson from Portfolio Analysis

One market-based principle of risk management that has an important bearing on questions of sustainability is portfolio balancing. Portfolio analysis indicates that the overall risk of a given energy choice can only be evaluated as a whole and not as the sum of its parts. In a portfolio with favourable risk qualities the risks of different elements can offset each other. For instance, suppose that one option (e.g., diesel-based generators) is profitable with a low oil price and another (e.g., a wind farm) is profitable if the oil price is high. Decision-makers interested in managing risk would then opt for building smaller versions of both, rather than depending on either one alone. This holds even if one option, say the generator, is on its own clearly less risky than the other, the wind farm.

Energy diversification has long been advocated by the IEA. Diversification has important implications for the development of new technologies such as renewables or energy efficiency improvements. One of the developers of the modern treatment of risk, Robert Lind, has said[39]*:*

"The development of an energy technology with very uncertain future returns may not constitute a risky project. If it will have a high payoff under just those conditions when the rest of the economy will do poorly, it will reduce the overall variability of national income and therefore will reduce risk. Such an investment has the characteristics of insurance."

In the context of sustainable development this interesting result can be taken one step further. In order to increase the resilience of the energy system, the diversification of energy choices should be pursued not only in relation to risks in the economic dimension, but also in the environmental and the social dimensions.

39. *Robert C. Lind (1982),* Discounting for Time and Risk in Energy Policy, *p. 15. Cited according to Ronald J. Sutherland (1991),* Market Barriers to Energy-Efficiency Investments, *p. 30. See also Shimon Awerbuch (2000),* Getting It Right: The Application of Portfolio Theory to Renewables Standards *for a more recent discussion in the context of renewable energies.*

affairs: recognising those [circumstances] from which evils arise – which is only given to those with great foresight – they can be quickly remedied; but when they are unrecognised and left to grow such that everybody will recognise them there will be no more remedy[40].

Discount Rates

Decisions about insurance and other types of long-term investment are strongly influenced by the discounting of future streams of costs and benefits. Since we have emphasised that policy-making for sustainable development is intrinsically oriented toward the future, it is useful to review this decision technique here.

The discount rate converts future values into present values. The higher the discount rate, the less the concern for the future. The classical formulation of the theory is based on the assumption of a preference for present consumption over future consumption and on the increase in wellbeing derived from economic growth. If the economy is expected to grow more quickly, I can allow myself higher consumption today.

The stronger the concern for the wellbeing of future generations, the lower will be the discount rate used by the decision-maker. Given the potential for disagreement on such an issue, it is not surprising that discount rates constitute a highly controversial subject. The energy sector is particularly subject to these disagreements. The high capital costs of energy projects and their long lifetimes often make the setting of the discount rate a determining factor in the choice between energy options (see box 3).

One controversy surrounding discount rates concerns the distinction between the private and the social rate of discount. It is argued that the social discount rate, used in government policy decisions, should be lower than the rate used in private decisions. The role of government as the guarantor of the public good has always included an obligation to preserve the long-run framework of the economy and society. This obligation has been translated in

40. *Translated from Niccolò Machiavelli (1515),* Il principe, *Firenze, Capitolo III "De' principati misti".*

Box 3

The Discount Rate and the Choice of Energy Systems

The IEA/NEA publication Projected Costs of Generating Electricity: Update 1998 *(Paris: IEA, 1998) compares the average per kWh costs of different energy options in different countries. The results are instructive. At a 5 per cent discount rate, gas and coal were the cheapest source in each of three countries and nuclear energy was the least expensive in five countries. At a 10 per cent discount rate, gas was the least expensive option in nine countries and coal in one country. Discounting at 10 per cent, no country would choose nuclear energy with its high up-front capital investment costs, which can only be amortised over very long periods.*

These outcomes are based on direct costs. The importance of the discount rate increases if the environmental effects of the different technologies are taken into account. Nuclear storage and greenhouse gas emissions are phenomena with potential effects far into the future. When taking them into account, the choice of discount rate can have an even more dramatic influence on the comparative evaluation of different energy technologies.

recent years into a commitment to sustainable development and more attention to the welfare of future generations – a commitment occasionally referred to as the "evolutionary imperative". On the other hand, the private commercial considerations that are expressed in market discount rates do not place much value on long-term effects. In the energy sector there is more experience with and acceptance of the need for long-term planning than in many other parts of the economy. Nevertheless there is little private value to investments that do not provide a net pay-off for the first 100 years. Some economists have opposed the concept of a social discount rate for certain investments, arguing that using different discount rates for different investments would yield economic inefficiencies. They have also made the point that reducing discount rates in general would raise the global level of investment, and hence the consumption of future generations, at the expense of current consumption even though future generations are likely to be richer.

Some environmental economists argue that discounting is not the issue but substitutability is[41]. Discount rates are based on the comparability and substitutability between different assets through time. They are useful only if comparative cost-benefit calculations can be made between different courses of action. Policymaking for sustainable development, however, requires making decisions on issues that are unique and non-substitutable. For example, suppose that global warming wreaked havoc on the global environment on a scale not commensurable with marginal increases in economic wealth. In that case, the assumption of substitutability breaks down and cost-benefit analysis based on discounting is no longer of any use.

One practical consequence of these considerations could be to leave private discounting procedures in place, provided that the economic values assigned to environmental assets that cannot be reproduced or substituted grow over time at a rate close to the discount rate. Another possibility would be that discount rates would decline over time – thus reflecting the risk that economic growth itself could slow in the future as a consequence of rising environmental costs[42].

It is obvious that some position must be found between *laissez faire* and the incurring of infinite costs to avoid climate change. Discount rates are a practical tool with which to set priorities in everyday decisions. Taking climate change as an example, once again, if there are no public incentives to square long-term emissions objectives and near-term private investment decisions, new coal-fired power plants with 50 or more years of expected

---

41. For a careful discussion of the connection between the two issues see Eric Neumayer (1999), Global warming: discounting is not the issue, but substitutability is.

42. See Cédric Philibert (2000), The economics of climate change and the theory of discounting *for a good overview of the literature on this subject.*

operating life will continue to be built in spite of their projected long-term contribution to CO_2 emissions. Bridging the gap between long-term social and short-term private discount rates is thus one function of government in the pursuit of sustainable development. Setting discount rates is ultimately a political rather than an analytical decision. Market interest rates on long-term bonds can give some guidance for marketable assets, but the treatment of non-replaceable assets will require the use of additional consensus-forming processes. This is particularly true for threats to sustainability at the systemic level, such as climate change.

Mastering Complexity: the Importance of Research, Learning and Informed Decision-making

Uncertainty and Complexity

The need for risk management is the expression of a still broader phenomenon – policy-makers have to respond to ever more complex challenges in ever more uncertain environments. This is especially true in the energy sector. Energy production and consumption interact with other phenomena in many ways, frequently in indirect and unexpected forms and a multitude of decision-makers and stakeholders are involved. Interdependence increases the complexity of the effects of each decision. Rapid structural change in the energy sector, in the form of market liberalisation, and elsewhere, such as moves toward a service-based economy and electronic commerce add further to the complexity. Global interdependence, an increased emphasis on the social and distributional consequences of energy policies, and an intensified orientation toward the long-run future also complicate the picture.

This combination of massive complexity and uncertainty enhances the importance of two kinds of strategies. First, because uncertainty can be reduced and complexity better dealt with through better understanding, conscious strategies are needed to support learning and research relating to sustainable development. Second, because uncertainty and complexity together make policy decisions more difficult, new attention to the processes of decision-making is needed.

Research and Learning

Research areas that will allow the formulation and implementation of better sustainable development policy in the energy sector include: energy supply and demand forecasting, the search for strategies to reduce greenhouse gas emissions, gauging the needs and prerogatives of oil producers and consumers, and determining effective strategies for the development of new technologies. Perhaps most interesting are those areas where increased knowledge promises progress in more than one dimension of sustainable development. This is true, for instance, of work on energy efficiency and renewable energy sources. Examples of recent activities are:

Box 4

Complexity and the Car

The personal automotive vehicle – the car – is at the heart of a large distinctly identifiable cluster of economic, technological, social and environmental phenomena. It involves several industries, including car manufacture itself, steel production, tires and rubber manufacture and the building of traffic infrastructure as well as a distinct lifestyle. It affects production, consumption and leisure patterns in a way that will ensure its central position for a long time to come.

The extent and complexity of this economic, technological and social cluster are such that transport policy is one of the most difficult and frustrating areas in energy policy making. Due to the strength of its internal linkages and the flexibility of many of its components, transport has remained almost impervious to public attempts to manage it. The cluster is in itself large enough to be subject to Say's Law that supply creates its own demand. It thus gains added weight through its implications for employment and general economic development. Henry Ford's commercial genius presaged this when in 1923 he paid his workers a wage high enough to allow them to buy a Model-T with 58 days wages[43]*.*

In the economic dimension alone, the "car cluster" is highly sustainable, though it severely tests the limits of sustainability in other dimensions. Environmental pollution, congestion, road deaths and oil security are the most important challenges that the transport system poses in regard to environmental and social sustainability[44]*. Matching the clout and complexity of the transport juggernaut would require developing dedicated policies that are equally broad, and applicable to economic, social and technological options (see Chapter 7).*

The co-development of economic, technological and social systems in clusters is, of course, not confined to the transport sector. The great steel producing centres of the 19th *century that were at the heart of the early industrial revolution provide another illustration. They developed close to large coal deposits in Manchester, the Ruhr area, Silesia, Pittsburgh and subsequently became manufacturing centres, exploiting closeness of transport links, cheap energy and proximity of demand. Large-scale migration, urban development, the formation of social classes and modern life-styles all developed contemporaneously.*

43. *McNeill (2000),* op. cit., *p. 316.*

44. *Arguably, the personal computer is at the center of another cluster that might eventually deeply affect the transport sector. The ambivalence of this prospect for sustainability can be somewhat incompletely summed up in the catch phrase "goods will travel instead of people". Whether it can challenge the "car cluster" as we know it in the foreseeable future remains to be seen.*

- New projects in the building sector on insulation, more efficient space heating and cooling equipment, building standards and housing retrofit programmes;
- Outreach to building trades, finance, energy and housing policy officials to make improved energy efficiency standard practice in building, renovating and marketing homes;
- New projects in the manufacturing sector focusing on industry audits, best practices and voluntary agreement programmes.

In the context of reaping the returns from research, "learning" refers to the process by which the cost of producing and using new technologies is reduced as experience with them accumulates and adaptations are made. Investments in learning, as well as in more traditional research activities, are an important part of what governments can do in support of developing markets for new energy technologies (This issue is discussed further in the context of renewable energies in Chapter 6).

The Need for Better Decision-making

New challenges, new complexities and rapid structural change raise questions about the decision-making capacities of existing institutions in the energy sector in relation to the goal of sustainable development. On the one hand, increased market orientation decentralises decision-making. On the other, increasing inter-connectedness requires new forms of global and regional co-ordination. Energy choices in fast-growing developing countries interact in a myriad of ways — global energy demand, supply security, energy-related greenhouse gas emissions, technology transfer and investment — with those of the OECD countries. International environmental efforts like the "Rio Process" and the UNFCCC have increasing influence on national policy-making. The role of regional bodies, such as the European Commission, rivals or even surpasses that of national governments in certain areas.

This new vertical stratification in decision-making and influence is being matched by more diffuse forms of horizontal co-operation.

Civil society, in the form of businesses and NGOs, is ever more involved in policy-making. Traditional policy-making by national governments is still important but developing clear strategies in the face of conflicting demands is increasingly a struggle.

What is needed is not necessarily the involvement of many individual actors at each step of the process, but rather the appropriate channelling of information from all different viewpoints, the establishment of scope for recourse and the building of consensus[45]. Policy integration also requires new institutional arrangements inside individual governments. Creating such new institutional arrangements can lead to overlapping competencies and confused signals during adaptation periods, an issue that has particular relevance in the area of climate change.

The need to broaden traditional hierarchic decision-making structures is driven both by the multi-dimensionality of sustainable development and by external developments. One result is that the traditional division of labour between governments and private actors is becoming blurred[46]. Governments today are expected to function much more like private companies – rewarding effort, performing to benchmarks, being accountable. Private companies are more and more often pushed to adhere to environmental, ethical and social standards. The increasing role of non-governmental organisations, the media and public opinion in influencing both constituencies is patent.

While some of these processes are driven by genuinely new social concerns and demands, the role of technology, in particular information and communication technology, in these processes is critical. Although visions of direct electronic democracy are probably premature, it is clear that technology is influencing the manner in which public opinion is formed. New possibilities for generating information widely at very low cost almost automatically make for broader stakeholder involvement and more diffuse decision-making structures.

---

45. Interestingly, establishing a consensus on concrete policy steps is often easier than creating consensus about ultimate policy objectives or the implications of scientific analysis. This experience has led the Dutch government to adjust some of the processes that feed into its policy decisions in the climate change area.

46. See, for example, Wealth, Values, Institutions: Trends in Government and Governance, *Report of the Special Advisory Unit to the OECD Secretary General.*

For several key issues the question of which institutional arrangements will best serve the objective of sustainable development in the energy sector is crucial. The quest for politically sustainable decisions in the context of risk, complexity and uncertainty requires broad democratic decision-making processes that (a) are robust even in the face of unpalatable events and (b) that integrate available information about effects, causes and preferences to the largest possible extent. These processes need to be open, transparent and flexible in order to react to new information and changing environments.

Box 5

Addressing Public Concerns

The insights of the Nuclear Energy Agency on involving the public in nuclear energy decisions apply as well to other energy sources that have large-scale effects on the different dimensions of sustainable development.

"Addressing public concerns is essential to meet the social objectives of sustainable development. For this purpose and in the light of the widespread public concern about nuclear risks, it is necessary to include the public in a democratic decision-making process through which it gains confidence that its concerns are being heard and addressed. The implementation of nuclear energy projects requires a participation of the public at the national and local level, and the exchange of a broad range of information and perceptions covering scientific, technical, economic and social aspects. It is important to allow the public to put social, ethical and political issues related to nuclear energy into perspective with the issues raised by alternatives, including liabilities passed to future generations such as geological disposal, climate change and resource exhaustion. It is the responsibility of governments to create the conditions for the decision making process to be consistent with the inter-generation equity and social objectives of sustainable development as well as with environmental protection goals[47]."

47. *Nuclear Energy Agency (2000),* Nuclear energy in a sustainable development perspective, *p. 5.*

The information needed for such processes includes objective, technical and scientific information as well as subjective preferences concerning exposure to different kinds of risk and the willingness to pay for their reduction. The generation, integration and translation of this diverse information into social preferences and political choices require the co-ordinating role of government. Despite the trend toward increased market liberalisation, energy choices remain political choices. However, in response to the challenge of sustainable development there is also a tendency toward a new division of labour in which other stakeholders participate, providing information, expressing preferences and shouldering some of the responsibilities. Caught between these two tendencies governments are slowly adapting to the new demands they are facing.

New institutional approaches to decision-making exist not only in the blueprints of experts on governance – they are increasingly becoming a reality. Sweden, for instance, has taken a very deliberate approach to designing an institutional structure to deal with sustainable development issues. It emphasises research and study, consultation and collective decision-making, followed by decentralised implementation, as well as by public information and education. The Swedish Ministry of the Environment, which is responsible for sustainable development, is rather small, but supervises and co-ordinates the work of thirteen policy-implementing agencies. The most important among these is the Environmental Protection Agency[48]. An Environmental Advisory Council assists in developing the overall policy orientation of the government. Ultimately, local authorities have a key role in the implementation of sustainable development through "green procurement" and the Local Investment Programme, which is funded with SKr 6.5 billion until 2003.

The Environmental Protection Agency encountered problems with policy coherence when it published detailed lists of environmental

48. The other twelve are the National Board of Housing, Building and Planning; Council for Building Research; Swedish Centre for Ecological Sustainability; National Chemicals Inspectorate; Governing Body of the Nuclear Waste Fund; National Land Survey; National Organisation for Aid to Owners of Private Small Houses; Swedish Geo-technical Institute; Swedish Nuclear Power Inspectorate; National Radiation Protection Institute; National Water Supply and Sewage Board; and the Swedish Meteorological and Hydrological Institute.

objectives. A new approach now concentrates on 15 key objectives endorsed by Parliament. Some objectives are highly ambitious, such as the target of CO_2 emissions of 4-4.5 tonnes *per capita* per year by 2050. But they do provide a clear direction for environmental initiatives, including the introduction of a carbon tax and a pioneering role in developing emissions reduction projects in developing countries.

Stakeholders were thoroughly involved in the decision of the Swedish parliament to phase out nuclear power. Nuclear power is a critical factor in at least three areas of sustainable energy development: the diversity and security of energy supply, the reduction of greenhouse gas emissions and risks from the production and long-term waste storage of nuclear power plants. The complex trade-offs that needed to be made were embedded in careful discussions and various accompanying measures, such as a SKr 9.2 billion programme to develop renewable energy sources and improve energy efficiency. Flexibility in the timing of the phase-out was also maintained. Regardless of whether the Swedish decision to phase out nuclear power was correct, it nevertheless provides a rich example of decision-making for sustainable development.

The Indicator Issue

Agreement on appropriate indicators for progress toward sustainable development is one of the first tasks that policy makers must face if they want their commitment to be credible. But, the relevant indicators and the links between them are not always obvious. The problem is compounded if the objective is to find aggregate indicators that synthesise information over several lower-level indicators.

Ultimately the indicators selected will vary depending on the purpose and level of analysis. The most common indicator to measure welfare, GDP, has clear shortcomings. To some extent the current concern about sustainable development expresses disenchantment with an over-emphasis on economic growth as a criterion for welfare and policy. At the same time, alternative measures, such as "genuine savings", face conceptual problems and are not widely accepted. That leaves disaggregated indicators for specific questions as the most likely choice.

In the energy sector, there is no one single indicator that can be used to assess movement along a sustainable path. Different indicators can, however, indicate movement toward sub-goals. The most obvious of these are energy production and consumption, as well as CO_2 emissions. Data on these and related items provided by the IEA Statistics Division remain the bedrock of quantitative analysis on energy.

It is useful to consider which policy-oriented indicators can help to describe the links between energy use and human activity in a way that provides a more complete picture than aggregates such as the ratio of energy use to GDP. The table below provides, as examples only, a selection of indicators. They illustrate the breadth and multi-dimensionality of possible indicators of sustainability in the energy sector.

In order to be useful for policy formulation at the national level, such indicators need either to be complemented by other more detailed indicators or to be interpreted in the light of national circumstances. The OECD *Analytic Report on Sustainable Development* highlights this trade-off in its chapter on indicators:

Although developing a set of indicators enhances coverage compared to single indicator approaches and allows increased adaptability to different national contexts, the use of a large range of indicators can make it difficult to communicate a sustainable development picture. In smaller, headline sets or dashboard models [broadly applicable indicators on selected issues] some degree of interpretation is required to compare and contrast different movements in different indicators but the limited number of indicators makes painting a sustainable development picture more straightforward[49].

49. *OECD (2000),* The Measurement of Sustainable Development, *p. 19.*

Table 4

Examples of Indicators in the Energy Sector with Relevance for Sustainable Development[50]

Economic indicators	
Average subsidy per effective unit of energy	Average of the unit subsidy for each energy source weighted by its share in energy consumption.
Consumption	Per capita consumption of final energy
Energy supply indicators	
Reliability	Percentage of time that a particular energy source is available for use (with back up where available).
Import dependency	Oil import dependency from particular countries or regions
Energy diversification	The sum of the squares of the shares of different energy sources in effective energy consumption.
Environmental indicators	
Greenhouse gases	Per capita emissions of greenhouse gases expressed in CO_2 equivalents.
Local emissions	Deposits of SO_2 per square kilometre
Social indicators	
Affordability	Ratio of a household's per capita effective energy consumption to a subsistence threshold.
Education	Hours of lighting available to schoolchildren
Health	Proportion of population affected by energy-related health problems such as respiratory illnesses.

50. *Based on World Bank (2000),* op. cit., *p. 39 and IEA Energy Statistics Division.*

All indicators are, by necessity, backward looking. They need to be carefully distinguished from forward-looking policy recommendations. By definition, an indicator isolates information about a single parameter in order to trace a measurable development through time. Obviously, the context from which the parameter receives its meaning will change over time, while the same parameter continues to be measured. And policy recommendations are context-dependent. For instance, increases and decreases in energy consumption have different policy implications in developing countries and developed countries. No amount of diligent indicator work can substitute for policy discussions.

An alternative to the establishment of quantitative indicators is, of course, descriptions of states of the world in qualitative terms. This is the approach that has been taken in the scenarios of the *World Energy Assessment* (see table 5). While this approach allows reflecting more illustrative detail it is less transparent and comparable. In the end, indicator work cannot escape the tension between these opposites – a conscious compromise has to be struck.

The best strategy might well be to rely from the start on a set of simple, highly transparent indicators that throw light on isolated aspects of sustainable development, without claiming to capture the whole. This is the approach taken in the chapter on indicators in the analytic report of the OECD's three-year horizontal project on sustainable development. In order to combine policy relevance with transparency, the OECD report concentrates on indicators of relevant human activities such as intensity of water and energy use or waste generation rather than on the measurement of biophysical resource stocks such as the extent of protected areas, forest cover, or resource exhaustion. In the framework of the standard pressure-state-response model, the OECD thus confines itself to the issue of state indicators (see table 6). This allows the assessment of progress against established policy objectives such as CO_2 emissions targets or education levels. It confidently assumes that such targets can be formulated without underlying information on environmental or social pressures like global warming or the crime rate.

Table 5

Characteristics of Sustainability in the WEA Scenarios[51]

Indicators of sustainability	1990	Scenario A3	Scenario B	Scenario C1
Eradicating poverty	Low	Very High	Medium	Very High
Reducing relative income gaps	Low	High	Medium	Very High
Providing universal access to energy	Low	Very High	High	Very High
Increasing affordability of energy	Low	High	Medium	Very High
Reducing adverse health impacts	Medium	Very High	High	Very High
Reducing air pollution	Medium	Very high	High	Very High
Limiting long-lived radio-nuclides	Medium	Very Low	Very Low	High
Limiting toxic materials	Medium	High	Low	High
Limiting GHG emissions	Low	High	Low	Very High
Raising indigenous energy use	Medium	High	Low	Very High
Improving supply efficiency	Medium	Very High	High	Very High
Increasing end-use efficiency	Low	High	Medium	Very High
Accelerating technology diffusion	Low	Very High	Medium	Medium

51. UNDP, UNDESA and WEC (2000), op. cit., p. 30. Scenario A3 depicts a high-growth, non-fossil future, scenario B middle growth and scenario C1 an ecologically driven new renewables future.

Table 6

A Preliminary Set of OECD Sustainable Development Indicators[52]

Theme	Current indicator
Resources	
Produced assets	Net value of capital stock
Financial assets	Net foreign assets
Environmental assets:	
Energy resources	Aggregate energy intensity index
Water resources	Intensity of water use
Climate change	CO_2 emissions from energy use
Waste	Municipal waste generation
Biodiversity	Percentage of threatened species
Human capital:	
Education	Per cent of population with secondary or tertiary schooling
Population growth	Fertility rates: total and completed
Labour force	Employment – dependency ratios
Technological change	Multi-factor productivity growth rate
Outcomes	
Income	Net national income (NNI) per capita
Income distribution	Ratio between top and bottom tenth
Health	Disability-free life expectancy
Work status	Standardised unemployment rate
Environmental quality	(to be determined)

The elusive idea of having one overall indicator for sustainable development continues to exert a strong attraction and has spawned numerous research efforts, notably at the World Bank[53]. Not the least of the problems of such a synthetic indicator is to

52. Ibid., 22.

53. *The following sub-chapter is largely based on Hamilton and Clemens (1998),* op. cit. *and Arundhati Kunte, Kirk Hamilton, John Dixon and Michael Clemens (1998),* Estimating National Wealth: Methodology and Results. *See also David W. Pearce, Kirk Hamilton and G. Atkinson (1996),* Measuring Sustainable Development: Progress on Indicators *for an early exposition.*

differentiate it clearly from traditional measures of welfare, such as GDP, without completely abandoning them. Economic wealth expressed in GDP is still an important measure of wellbeing, even if most analysts would agree that it is an insufficient indicator of total welfare.

The indicator of "genuine savings" has received much attention for its grounding in economic theory and compatibility with the familiar GDP indicator. "Genuine savings" expands the elements that are included in the definition of GDP[54]. In addition to GDP, it includes physical capital formation (investment), the depletion of natural resources, environmental pollution and human capital formation (education). The intent is to provide a concrete measurement of weak sustainability by adding natural and human resources to net national product.

In developing policy-relevant estimates of "genuine savings", a number of controversial practical issues arise:

- Measurements of pollution damages, or environmental externalities, frequently do not exist. Where they do exist, they can involve wide ranges of uncertainty.
- The link between educational expenditures and human capital formation is tenuous.
- Even the measurement of natural resource depletion, the component of "genuine savings" where cost should be most easily estimated, involves a large number of controversial conceptual issues: marginal vs. average cost, the treatment of quality differences and wide price swings.

Cross-country comparisons of "genuine savings" rates seem biased against resource-exporting economies and in favour of high-growth, resource-poor economies. World Bank data for 1993 indicate that Korea, Japan and Singapore are sustainability leaders, while Venezuela and Saudi Arabia are lagging behind. Among OECD

54. Strictly speaking, "genuine savings" expands upon net national product (NNP, which is equal to annual consumption plus changes in the capital stock plus changes in the net foreign asset position) taking account of resource depletion, pollution and education expenditures.

countries, Canada, Finland, Norway and the United Kingdom also fare comparatively badly.

The results from calculating genuine-savings rates contrast starkly with those obtained in the *2001 Environmental Sustainability Index,* a joint effort of the World Economic Forum, Columbia University and Yale University[55]. The Environmental Sustainability Index (ESI) is composed of 22 indicators in the following five categories: environmental systems (5), reducing stresses (5), reducing human vulnerability (2), social and institutional capacity (7) and "global stewardship" (3).

The emphasis in the ESI approach on health-related environmental issues, especially water and air quality, and on institutional capacity favours high-income countries with low population densities. The ESI is very closely correlated to *per capita* incomes, although the authors emphasise that countries with similar incomes can have quite significant differences in their ESIs. Finland, Norway, Canada and Sweden turn out to be the four countries with the highest scores under the ESI methodology, whereas the low score of Singapore (65 out of a total of 122 countries surveyed) led to an official complaint from the country. Thus far, the ESI and the genuine-savings index give dramatically different results. It is also interesting to note that under the dynamic genuine savings approach the exploitation of renewable natural resources was penalised, while these very same natural resources contribute to good environmental quality in the static ESI approach.

These two examples show that efforts to find new measures of national wealth accounting for purposes of sustainable development still have some way to go[56]. In some ways, *ad hoc* measures such as the Human Development Indicator (HDI), which simply combines GDP *per capita,* life expectancy and adult literacy into one indicator, seem more relevant than "genuine savings" or the ESI to the policy

55. *World Economic Forum, Yale University and Columbia University (2001),* 2001 Environmental Sustainability Index.

56. *See Kirk Hamilton and Ernst Lutz (1996),* Green National Accounts: Policy Uses and Empirical Experiences *for a history of experiences with a number of concepts preceding "genuine savings".*

concerns surrounding sustainable development[57]. From an energy point of view, it would be interesting to add "per cent of population with access to electricity" as a further element in the HDI.

A single ideal solution to the need for indicators of sustainable development does not exist. There are several options, each of which is more or less useful according to the context. Sustainable development remains a moving target in a highly complex environment. Quantitative indicators are only one instrument with which to define the framework conditions in which actions are to be taken to manage risks and to balance the different dimensions of sustainable development in the energy sector.

57. *The Human Development Indicator is published annually as part of the* Human Development Report *by the United Nations Development Programme (UNDP). For a complete country by country overview of the latest HDI see http://www.undp.org/hdro/98hdi1.htm.*

SAFEGUARDING ENERGY SUPPLY SECURITY

Introduction

Maintaining a stable supply of energy is a core objective of policy in the pursuit of sustainable development. Its importance for economic and social sustainability is paramount[58]. Energy supply interruptions constitute the sort of systemic risk that needs to be addressed by policies for sustainable development. This requires addressing the sources of risk as well as the potential impacts – economic and social disruptions due to inflexibility in consumer behaviour. The continuous management of energy supply risk is an integral part of sustainable development.

The costs of supply disruptions go beyond the abstract language of national economic accounts. Energy use pervades daily life in such a constant and ubiquitous way that it is often taken for granted, which makes the shock of its sudden absence all the harder to bear. The empty streets, the dramatic exhortations to save energy and the ensuing stagflation of the oil crises of 1973 and 1979 have durably marked collective memory. More recently, the power supply crisis in California has reminded us that disruptions can happen again, in unexpected ways.

Safeguarding energy supply security has a short-term and a long-term component. In the short run, emergency measures such as the co-ordinated use of energy stocks, temporary blackouts and re-directed supply flows can help to minimise economic disruptions by allocating scarce supplies where they are most needed. In the medium and long run, policies to increase energy efficiency, diversification of fuel sources and reserve margins can limit the general exposure to disruptions and enhance flexibility.

58. In regard to the third dimension of sustainable development, the environment, energy security has only secondary importance. To the extent that energy supply interruptions lead to reduced energy consumption, they may even reduce environmental impacts from energy. However, this effect is minor in comparison with their potentially dramatic impacts on economic and social sustainability.

The current situation in IEA/OECD countries requires heightened attention to supply security issues. The latest *World Energy Outlook* projects a steady upward trend in oil consumption in the OECD area and globally over the next decade, underpinned by growth in the transport sector. With domestic oil production forecast to decline, net oil imports are expected to rise faster than consumption. While IEA countries' dependence on net oil imports fell from about 70 per cent in the mid-1970s to about 50 per cent in the mid-1980s, it has increased steadily since then and is likely to exceed 70 per cent again in the next decade.

Although the risks of macroeconomic damage from high oil prices may have diminished somewhat, the geopolitical risks of supply disruption remain. The bulk of future incremental oil supply is expected to come from the Middle East. Moreover, the scope for sudden increases in production and fuel switching in IEA/OECD countries is limited and demand restraint may be less effective as a response measure than in the past. An increasing portion of oil demand is in a transport sector that is seemingly impervious to efforts to constrain it through policy (see also Chapter 7).

What is an Energy Supply Disruption?

Energy security is defined in terms of the physical availability of supplies to satisfy demand at a given price. The security problem therefore involves a quantity risk and a price risk. It also has a long-term and short-term component: a long-term trend of, say, rising prices for energy imports has a different implication for an economy than a sudden price hike or price volatility. The first difficulty in addressing energy security is defining the nature of the problem in these terms and attempting to evaluate the costs of failing to meet security objectives. It is important to understand, however, that the societal cost of a primary energy supply crisis tends to be heavily influenced by the specific historical setting and the geopolitical situation in which it occurs.

Oil supply disruptions have occurred rather frequently: over the past half century there have been at least 14 significant disruptions involving a loss of 0.5 mb/d or more of crude oil. Most of these disruptions were related to political or military upheavals, especially in the Middle East. Since 1973, four major crises – the 1973 Arab-Israeli War, the 1978/9 Iranian Revolution, the 1980 Iran-Iraq War and the 1990-91 Gulf War – resulted in initial shortfalls of between 3.0 and 5.6 mb/d. Virtually all past oil disruptions have been short, typically lasting no more than six to nine months.

The politically-based disruptions can be broadly grouped into two categories:

- Random shocks, caused by internal unrest in OPEC countries such as the Iranian Revolution in 1978/79, the Nigerian civil war of 1967-70 or wars involving OPEC states like the Iran/Iraq War and the 1990 Iraq invasion of Kuwait.
- Strategic shocks involving wilful exercise of market power by Middle Eastern oil producers such as the Arab oil embargoes of 1957, 1967 and 1973-4.

It has been found that medium-term physical unavailability during oil supply interruptions is less likely in OECD countries if prices are allowed to adjust. Fuel substitution takes place where it is possible to replace "missing" fuel supplies associated with a disruption. An IEA study on the security of natural gas supply investigated the effects of sudden, short-term supply disruptions such as those associated with political instability in Algeria and Russia, dramatic cold spells in North American or sudden, unforeseen inoperability of one or several re-gasification terminals in Japan[59]. One conclusion was that this type of supply disruption results in substitution by other supplies or other resources in the medium, or even short term. For example, the unavailability of natural gas for power generation has led to more use of light distillates in combined-cycle gas turbines, heavy fuel oil in oil plants, fuel switching to coal and oil in multi-fired

59. *IEA (1995),* Natural Gas Security Study.

Table 7

World Oil Supply Disruptions

Dates	Supply Disruption	Magnitude of Supply Shortfall (mb/d)	World Oil Consumption (mb/d)	Per cent of Consumption
Mar. 1951 - Oct. 1954	Iranian Fields Nationalised	0.7	13.2	5.3
Nov. 1956 – Mar. 1957	Suez War	2.0	17.5	11.4
Dec. 1966 – Mar. 1967	Syrian Transit Fee Dispute	0.7	34.3	2.0
Jun. 1967 – Aug. 1967	Six Day War	2.0	40.0	5.0
Jul. 1967 – Oct. 1968	Nigerian Civil War	0.5	40.1	0.3
May 1970 – Jan. 1971	Lybian Price Controversy	1.3	48.0	2.7
Apr. 1971 – Aug. 1971	Algerian-French Nationalisation Struggle	0.6	50.2	1.2
Mar. 1973 – May 1973	Lebanese Political Conflict	0.5	58.2	0.9
Oct. 1973 – Mar 1974	October Arab-Israeli War	4.3	58.2	7.4
May 1977	Damage at Saudi Oilfield	0.7	62.1	1.1
Nov. 1978 – Apr. 1979	Iranian Revolution	5.6	65.1	8.6
Oct. 1980 – Jan. 1981	Outbreak of Iran-Iraq War	4.1	60.4	6.8
Mar. 1989 – Apr. 1989	Exxon Valdez Accident	<0.5	51.6	<1.0
Apr. 1989 – Jun. 1989	UK Cormorant Platform	0.5	51.6	1.0
Aug. 1990 – Jan; 1991	Iraqi Invasion of Kuwait	4.3	66.3	6.5

Source: US Department of Energy and the IEA Secretariat.

* *Initial production loss only: in some cases, this loss was quickly made up by production increases elsewhere.*

units or production increases in non-fossil plants in cases where there is spare capacity.

At the same time there are limits to short-term fuel substitution. Switching immediately to the next least expensive fuel might not be possible because equipment capable of using the substitute fuel may not be available or the supply infrastructure for the substitute fuel may not be adequately developed. Capital stock in energy markets is slow to change, mainly because of the comparatively long lifetimes of energy production, transportation and transformation equipment, and because of the lack of fungibility of certain assets. Substitution is limited by the availability of substitute fuels and adequate equipment.

Excess demand for a fuel that is not met by substitution will be quickly translated into price rises. More expensive supplies, end-users willing to forego the use of functioning sources of supply, and other sources of supply will appear as the price rises. Demand will naturally fall in response to these rises and market equilibrium will be re-established relatively quickly. In sum, a pure "quantity-risk" does not exist since physical unavailability of an energy source will lead to fuel substitution and price rises.

Box 6

The IEA and Energy Security

The mandate to safeguard energy supply security is at the origin of the founding of the IEA. The Agency works on both short-term and medium-to-long-term energy security. Concerning short-term supply security, the IEA's Energy Planning and Preparedness Division (EPPD) maintains a programme on oil emergency preparedness under the Standing Group on Emergency Questions (SEQ). These provisions were embodied in the International Energy Programme (IEP), the agreement by which the Agency was established in 1974.

The IEP commits IEA countries to reduce oil demand and share available oil in the event of major oil supply disruptions. It also requires participating countries to hold oil stocks equivalent to at least 90 days of net oil imports of the previous calendar year. An Industry Advisory Board composed of senior supply experts from international oil companies advises the SEQ on oil emergencies and related questions. Additional IEA measures to respond to oil supply disruptions include:

- *Early and co-ordinated use of stocks,*
- *Measures to reduce oil consumption,*
- *Short-term fuel switching,*
- *Increased indigenous production.*

IEA countries recognise that energy developments are of global importance and that sound energy policy actions both inside and outside the IEA would contribute strongly to improved energy sustainability. This provides the basis for the exchange of information with non-Member countries and for the IEA's extensive participation in the Consumer – Producer dialogue with oil-producing countries outside the OECD.

In the medium and long term, energy security is a question of energy efficiency, diversification and the maintenance of flexibility in the provision of energy services. The IEA's Energy Diversification Division studies the link between the diversified mix of major fuel sources, the availability of technical flexibility in the provision of energy services (e.g., dual-firing capacity) and the overall security of energy supply. Recent work includes study of the implications of energy market liberalisation on the energy mix and energy supply security.

The Nature and Magnitude of Supply Risk

Although supply disruptions are unlikely to bring all or part of an economy to a grinding halt, energy price increases can have major, economy-wide effects. The box below provides an overview of the potentially damaging effects on the economy of pronounced, unforeseen price increases.

Various studies have attempted to establish whether or not there are linkages between higher oil prices and reduced economic growth rates. At first, it was widely believed that the 1974-75 recession and the emergence of "stagflation" – a combination of rampant inflation and rising unemployment – in the industrialised countries was caused primarily by the oil price increases of 1973-74. However, later studies, which concentrated more on the role of fiscal and monetary policy, suggested that recessions in the oil importing countries were less the direct result of higher prices and more the consequence of the economic polices adopted to alleviate the price shock.

Bohi and Toman (1996) in a more recent overview point out that there is widespread controversy over the market power OPEC was able to exert during the oil crises. No econometric evidence shows whether the price increases reflect cartel rents, (i.e., the skimming-off of wealth due to market power), or unavoidable and economically acceptable scarcity rents due to strong oil demand growth combined with the time lag required to build up new supply capacity.

They also note that oil stocks increased during the first oil crisis, which might indicate that supplies were withdrawn from the market as prices rose, thus exacerbating scarcity through hoarding. Even during the second oil crisis, there was a fair amount of "irrational (panic) investment in the sense that inventories were purchased at high prices and sold at low prices" (Bohi, 1983), aggravating the economic burden from the original shortage and influencing market prices. In order to cope with such expectation-driven market volatility, the IEA established its Co-ordinated Emergency Response Measures, which emphasise early intervention using the co-ordinated use of stocks and other measures to calm markets[60].

The econometric evidence on the indirect effects of the oil price shocks is even less clear. Increasing prices of imported energy mean that the country under consideration must export more goods to buy the same amount of imports, that is its "terms of trade" deteriorate. Such changes in the terms of trade can lead quickly to deficits in the balance of payments and lead to a depreciation of the importing country's currency. However, even theoretically it is unclear whether an increase in energy resource prices must cause a decline in the importing country's exchange rate. This is because the exchange rate is determined by *relative* capital flows, i.e., it also depends on the exporting country's willingness to hold the importing countries' currencies. Empirical studies appear to show that the terms-of-trade effect of higher oil prices can be positive or negative for any oil-importing country, depending on the specific circumstances[61].

Inflation and labour market effects follow this pattern. While increased energy resource prices will raise prices in a one-off inflationary shock, *sustained* inflationary pressure requires an increase in the *growth rate* of oil prices – in other words, oil prices must increase faster and faster. An inflationary spiral as experienced in the 1970s requires additional explanations including excess demand, inflationary expectations and flawed monetary policies. Sustained mass unemployment may owe as much to labour market rigidities as to an external energy price shock.

Since the early 1970s no producing country outside OPEC has held production significantly below capacity – thus increasing the potential for supply disruptions. The difference between current production and capacity is estimated at 6 to 8 per cent of installed

60. IEA Governing Board decisions (1981 and 1984).

61. Marion and Svensson (1986).

Box 7

Price Changes in Response to Oil Supply Disruptions

A small number of supply/demand models capable of generating forecasts of near-term price movements (weeks or months ahead) or simulating supply disruptions provide some help in predicting price responses to supply crises, by determining a baseline around which prices will fluctuate according to expectation and speculative factors. For example, DRI/McGraw-Hill maintains a large-scale econometric model of the world spot oil market and produces regular near-term price forecasts and simulations for its clients. The U.S. Department of Energy (DOE) also maintains a spreadsheet model designed explicitly to predict oil price and demand responses in the event of a major supply disruption.

Based on the DRI model, a sudden cut in oil production of around 1 mb/d leads to approximately a US$ 1.50-2.00 per barrel average increase in crude oil prices within about three months, assuming prices in the US$ 15-20 per barrel range. In the DOE model, prices are more sensitive to supply losses. In broad terms, each 1 mb/d of lost supply typically increases spot crude oil prices by US$ 3.50-4.00 per barrel during the first quarter and progressively less for the following 2-3 quarters.

The lack of a clear correlation between initial supply losses and the resulting oil price increases occurs because during past crises those losses were in most instances rapidly offset by production increases in countries not affected by the crises. During the 1950s and 1960s the impact of supply losses was mitigated by the activation of shut-in production in the U.S. and elsewhere. In the U.S. alone spare capacity was nearly 3 mb/d in 1965 and 2 mb/d in 1969. However, by the beginning of the 1970s, this spare capacity had practically disappeared.

capacity which currently corresponds to less than 2 mb/d for non-OPEC suppliers an amount needed for contingencies such as maintenance.

During the 1960s OPEC countries behaved like any other producer by maintaining only the capacity margin needed for normal maintenance. However, since 1973 they have developed significant spare production capacity, initially by restricting production and subsequently by expanding capacity at a time of falling demand. As a result, a small number of OPEC countries, notably Saudi Arabia, have accounted for virtually all the spare production capacity since the early 1970s.

This brief discussion reveals the uncertainty of the potential costs to society attributable to energy resource price increases. Moreover, analysis of price movements in oil crisis situations reveals the important role played by expectation, speculation and market psychology, in addition to market fundamentals. Risk management in order to maintain continuous supplies and to ensure sustainable development thus requires a dual set of measures. A first set of measures is needed to focus on stable long-term fundamentals enhancing market transparency and trust between the major actors. A second set of measures is needed to prepare short-term responses for unforeseen crises in order to contain the fall-out from disruptions.

Demand Responses to Oil Price Shocks

There is a considerable body of literature on price and income elasticities of the demand for oil. These analyses shed light on how past price changes, particularly those resulting from oil shocks, and increases in real income have affected the level of demand for oil. The results can be used to predict future oil demand given assumptions about prices and economic growth within the framework of a model, such as the IEA's World Energy Model.

The changing structure of oil demand has had a major impact on price elasticities. There is reason to suppose that overall short-term price elasticity may have fallen significantly over the last decade for the following reasons:

- In the OECD, transport's share of total oil use has risen from 40 per cent to nearly 60 per cent over the past 20 years.

Transport demand may be the least sensitive of the main oil product categories to fuel price changes in the short-term. As a result overall oil product demand has probably become less price elastic.

- Taxation of oil products generally and transport fuels in particular has increased in many countries. This means that changes in crude oil prices have less impact on product selling prices, cushioning the demand impact of higher spot prices.
- Relatively elastic demand from industry and power generation today accounts for a relatively small share of total oil demand. The current potential for fuel switching in industry and power generation in cases where oil is still the preferred fuel is thought to be very limited across the OECD.

These factors suggest that spot crude oil prices might have to increase by more than in the past to reduce demand in any future prolonged supply disruption. In other words, the decreased flexibility or lower elasticity of oil demand increases the potential for severe economic disruptions from supply shortages.

Supply Risk as an Energy Security Externality

Externalities are costs or benefits related to the production and use of goods or services that are not borne by the producer or consumer. This concept, best known for its application in the environmental area, can also be applied to energy supply interruptions. An "energy security externality" associated with the use of a given fuel is the cost of an interruption in its supply that is not borne by the purchasers of the fuel.

With this definition, a security cost that energy consumers can take into account in their private decisions is not an externality. For instance, individuals or companies who buy a particular fuel may in various ways buy insurance against supply interruptions, however, since they bear this cost themselves it is not an externality. External costs are those that accrue to others in the economy. As an illustration, if a power company that uses imported oil incurs losses due to a supply interruption, these are internal costs. Moreover, if the company wishes, it can insure against such a risk and the cost of the insurance is an internal cost. However, if a school can no longer operate because the power company's supply of electricity is interrupted and the school cannot be properly heated, an external cost is also incurred. Because such external security costs exist, there is a need for governments to step in and take protective measures.

In theory an externality can be accounted for and allocated to its source; that is, it can be included in the price paid for the good that causes it. Doing so is said to "internalise" the externality. In the case of an energy security externality, this could mean that the costs for emergency response measures, diversification and other instruments to manage the risk of supply disruptions would be borne by the importers and eventually the consumers of the fuel that causes the externality. However, establishing the economy-wide cost of energy supply disruptions is fraught with difficulty. While it may be feasible to estimate the economic costs of a supply disruption or a sudden increase in price, it is not possible in practice to distinguish between internal and external costs in a comprehensive way.

Nevertheless attempts have been made to estimate the external portion of the costs of oil supply interruptions. These estimates vary widely. One study carried out by the U.S. Department of Energy estimates that an annualised external cost for the period 1990-2020 lies in the range of US$ 0.44-1.27 per barrel of oil consumed[62]. The estimate is reduced to US$ 0.17-0.49/bbl if the Strategic Petroleum Reserve is taken into account. Thus, according to these estimates, the oil security externality for the United States amounts to some 1-3 per cent of current U.S. crude oil spot prices. A recent study by the U.S. General Accounting Office supports these findings and states that further reductions in dependence on foreign oil imports would prove costly to the

62. *DOE (1990).*

economy while providing little additional protection from oil supply disruptions[63].

Supposing for the sake of argument that it is possible to come up with dependable estimates of energy security externalities, what is the best way to internalise them and who shall bear the costs of this internalisation? In the case of supply risk that arises because a particular energy source, say oil, is imported from another country, one approach is the application of an import tariff. This would allow substitution along the entire energy chain and would cause all consuming sectors to adjust to the measure. Electricity generation from oil would be reduced, people might decide to use their cars less often, freight might be shipped more frequently by train or barges and lubricants might be produced from non-mineral oils. If the tariffs were high enough, people might even start using natural gas or electric vehicles instead of petrol-fuelled cars.

While a particular concern has always existed about the security of oil supply for power generation due to the large externalities that can arise from an electricity supply interruption, for IEA/OECD countries in general this risk is limited. In 1999 oil accounted for only 7 per cent of total OECD electricity generation[64]. However, for a few countries with relatively large shares of petroleum in power generation, such as Italy, Portugal, or Japan, the risk is more critical.

Due to the overwhelming focus of research on oil supply and its potential disruptions, and the relative scarcity of studies on other primary fuels, the above discussion is incomplete. Ideally, the security costs of *all* primary fuels should be researched. Natural gas would especially deserve attention, as it is becoming the fuel of choice in countries that are liberalising their power markets. The 1984-85-coal miners' strike in the United Kingdom also suggests that even coal markets, which operate relatively smoothly, might give rise to some – albeit limited – supply risk.

---

63. GAO (1996).

64. See IEA (2000d), p. I.24f.

Diversity as a Means of Improving Energy Security

Optimal Plant Mix

Safeguarding supply security requires planning for all future contingencies, including the possibility of electricity supply disruptions. While there is currently no identifiable need for emergency measures in the power sector similar to those in the oil sector, long-term supply security calls for a balance between energy sources and technologies in order to reduce exposure to sudden problems. This focuses attention on diversity in supply. Diversity acts as insurance against various kinds of problems. For example, diversity in power plant technology reduces the risk of basic design flaws causing large numbers of plants to be shut down for repair or retrofitting, although it can come at the cost of reduced economies of scale. Recognising the advantages of diversity also raises the question of whether liberalised electricity markets may reduce it below an acceptable threshold.

The issue of an optimal plant mix in the presence of uncertainty is often discussed in the framework of portfolio theory, the framework used to determine an optimal mix of financial assets for an investor. For the financial investor the trade-off is between risky assets with high returns and safe assets with lower returns. In the case of power plant investment decisions, the trade-off is between low expected prices combined with a relatively high level of uncertainty and higher expected prices with a lower level of uncertainty. Adding some higher-cost generating options can act as an insurance policy against large price increases in fuels consumed in low-cost plants.

The discussion of discount rates in Chapter 2 is relevant in this context. It is argued that higher discount rates, which might be required by investors in more competitive power markets, reduce the incentive to invest in insurance against the risks of power disruptions by reducing the expected value of future electricity cost increases stemming from fossil fuel price increases.

A study based on portfolio analysis carried out for Scottish Nuclear implicitly suggests that competitive power markets provide less diversity[65]. The study argues that it is advantageous for society to insure itself against the risk of price increases from fossil fuels by opting for diversity, notably by using non-fossil energy, especially nuclear power, as insurance. The study argues that nuclear power significantly reduces risk at little extra cost. A portfolio with 30 per cent nuclear generation has an expected generation cost of 3.47 pence/kWh with a 1-in-90 chance of rising to 4 p/kWh. Generating costs of 4 p/kWh and above are defined as indicative of high cost in the study. In contrast, a portfolio with 4 per cent nuclear has average costs of 3.33 p/kWh, but with a 1-in-30 chance of rising to 4 p/kWh.

The first portfolio – the 30 per cent nuclear scenario – constitutes an optimal solution in the framework of the study. It is not much higher than generation from existing nuclear plant at the time the study was conducted, since in 1993, nuclear generation in the United Kingdom amounted to 27.8 per cent[66]. This basically means that maintaining existing nuclear capacity, and possibly adding one more reactor, would be an optimal strategy. In the longer run, when nuclear plants have reached the end of their technical and economic lifetime, maintaining this optimum would require new nuclear investment.

The results of the study were subjected to sensitivity analyses. The insurance value of nuclear power is positive if consumers are strongly risk-averse and still remains positive if they are assumed to be only moderately risk-averse. Risk-neutral consumers would in any case not be willing to pay an insurance premium. Risk-seeking behaviour was not modelled.

The central recommendation of the study is that the difference in cost between the mix of plants chosen by the market and the optimal portfolio solution should be borne by governments. The study acknowledges that the portfolio approach does not entirely fit the

---

65. *Scottish Nuclear (1994).*

66. *IEA (1994).*

reality of a competitive market. In a free market, there would be no such thing as overall cost and risk optimisation. Rather, all consumers would decide for themselves where they want to be positioned in the trade-off between current prices and price risk. In fact, it was argued that one of the drawbacks of centralised power systems is the lack of choice for final consumers in regard to risk, prices and reliability.

Unfortunately, the aggregation required for this kind of analysis poses some fundamental methodological problems. The authors argue that their results are meaningful as long as not all consumers have full access to competition, and as long as contractual arrangements that allow all customers to fully express their price/risk preferences are lacking. Drawing a comparison between the full opening of the United Kingdom electricity market in 1998 and fixed-interest arrangements in the mortgage market, they reckon it could take many years until all mechanisms, including intermediaries such as power marketing organisations, are built up, and until consumers themselves have gathered sufficient expertise to ensure that their preferences are optimally reflected in investment choices. This is disputable, since numerous energy firms and even supermarket chains positioned themselves to market electricity to residential consumers as of 1998.

As competitive electricity markets evolve, consumers and suppliers are likely to learn to take diversity into account and to develop coping mechanisms. Once the appropriate contractual arrangements are in place for everybody, the need for governments to alter the plant mix is vastly reduced. Only externality costs might justify government action in liberalised power markets.

Diversifying Fuel Supply Sources

While discussion in the preceding section was focused on diversification based on different input fuels, the diversification of supply sources of a given fuel is another option. The effectiveness of such diversification depends on the relationship between movements in the prices of the relevant fuels. Generally speaking, the less the prices of energy sources are linked to each other, the greater the value of a diversification strategy.

Where substitution between fuels is easy and fuel markets are competitive, the prices of different fuels per unit of energy content lie close to each other. This has traditionally been the case for oil and gas. Supply disruptions and the resulting price surge in one of the markets would thus quickly spill over into the other. In terms of microeconomic analysis, there is a high cross price elasticity between these markets. In econometric analysis this would be reflected in high covariances between prices. If a government believes that the existing fuel mix in the country's energy system is insufficient for energy security. This would increase the role of fuels or suppliers with prices that have had lower covariances with the prices of fuels already used. Coal might be one of these, as might be nuclear or renewable energies.

Market conditions, especially input fuel prices, are bound to change from time to time. Such changes could lead to more variety than the Scottish Nuclear study seems to suggest. And things *can* change in the energy market, even if such changes take some time to manifest themselves due to the long lifetimes of power plants, which, if depreciated, may be very successful in competing against new construction. During the "dash for gas" in the United Kingdom, the share of gas in power generation rose from 1.1 per cent in 1990 to 19.0 per cent in 1995 and to 32.4 per cent in 1998, with nuclear and coal roughly sharing the remaining two-thirds. The situation in 1998 was thus remarkably more diversified than in 1990, when two-thirds of all power was generated by coal[67].

Diversity Index

An index developed by Stirling helps to describe the level of diversity in electricity systems[68]. The Stirling index was developed on the premise that much of the uncertainty surrounding fuel

67. *IEA (2000d), p. II.664.*

68. *Stirling (1994).*

prices is ignorance, not risk. From this he concludes that it is inappropriate to apply portfolio analysis to electricity supply systems. In formal or theoretical terms, risk is generally defined in relation to a variable for which there is an objective method of assigning probabilities to possible outcomes. Uncertainty, or ignorance, occurs where such a mechanism is lacking, i.e., where the outcome is not predictable in terms of probabilities. Because fuel price movements depend on irreversible risk, such as unique foreign policy and military events, they are uncertain. Since portfolio analysis relies on reversible, hence measurable, risk, it is inappropriate in this case (see also the discussion on page 53).

Stirling instead analyses risk in electricity systems by applying a diversity index (DIV) borrowed from physics, the "Haynes' uncertainty measure", which is used in statistical mechanics as well as in the study of entropy in thermodynamics. The diversity index is defined as:

$$DIV = -\sum_i p_i \ln p_i$$

where p_i represents the proportion of fuel type i in an overall portfolio. The formula multiplies the proportion of each fuel type by its natural logarithm and then sums these values. Because the natural logarithm of a fraction is always negative, the minus sign at the beginning of the equation ensures that the index is positive. The index increases as the number of different supply sources increases. It is essentially a measure of the "evenness" of the plant mix. The index yields a lower value (less diversity) for five technologies that each have the same share of output, than for 50 technologies each with an equal share, and a higher value than for five technologies with an uneven share.

Based on government quantity and cost data and forecasts, Stirling calculates a hypothetical optimal electricity supply system for the United Kingdom. He uses an optimisation model that balances plant financial performance and the diversity of the plant portfolio as a whole, as expressed by the above index. As shown in the table below, the optimal result calculated with this model is very different from the system that existed in the United Kingdom at the time.

Table 8

Existing and Optimal United Kingdom Electricity Supply System (Per cent)

	Existing system	Stirling optimum
Nuclear	22	14
Coal	66	27
Oil	9	7
Gas	1	20
Renewables:		
Intermittent	0	13
Firm	0	17

Source: Stirling (1994).

The Stirling index has also been derived as a time series. Based again on UK government data, it leads to the diversity curve depicted in Figure 10. This graph shows increasing diversity in 1984 and 1985, when coal burning was reduced due to the miners' strike and when substitute oil use improved diversity. After the strike, and after falling back to its previous level, the index rose again in the late 1980s as coal began to lose ground to nuclear power, and subsequently to natural gas. It peaks in 2000, but declines again afterwards, when gas increasingly dominates the fuel mix.

The Stirling index is very controversial[69]. Most commentators agree that it has one large drawback: it treats diversity and "evenness" as desirable in their own right and makes no distinction between desirable and undesirable states of the world. The case of a hypothetical utility that has only coal and oil plants helps to illustrate this. To diversify, a third or fourth energy input should be used – natural gas, nuclear or renewables. Which one depends on which uncertainties one considers most serious: the lack of safe, long-term storage for radioactive waste, the potential security

69. See *Lucas* et al. *(1995), Brower (1995) and Stirling (1995).*

Figure 10

Diversity Index for UK Electricity Generation, 1984 to 2020

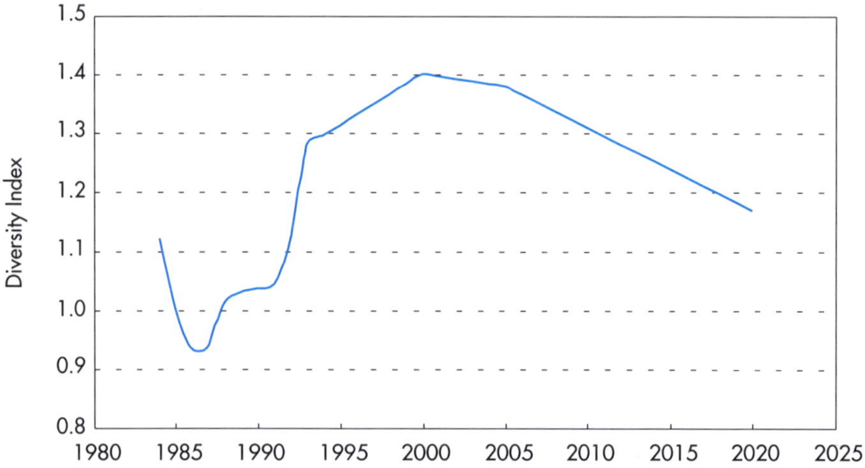

Source: Stirling (1994).

concerns associated with natural gas, or the uncertain costs and performance of renewables[70].

The Stirling index does not provide any guidance on the best choice in this sort of situation because it does not incorporate expected values of these variables. Its greatest strength, dispensing with the need to assume probability distributions for these variables, is also its greatest weakness. However, it may still be useful as a supplement to more conventional probability analysis.

The Value of Diversity in Electricity Generation

No matter how diversity is measured, the extra cost to society of more diversity in the power market still needs to be compared with the value of energy security externalities that this diversity is intended to reduce. The evaluation entails two parts:

70. It may seem strange to discuss fuel risk in relation to renewables, since they often have no fuel input or one that is purely domestic. Nevertheless, while fuel cost escalation is less of a problem, capacity cost overruns and under-utilisation can still cause large cost uncertainty.

- estimating the difference in electricity cost between plants intended to provide more diversity, and from plants that would be chosen otherwise;
- estimating the value of energy security externalities for various energy sources.

In practice this means comparing the cost of increasing diversity by using more non-fossil plants with the value of energy security obtained through this strategy[71]. If the energy security externality proves to be as large as or larger than the extra cost of increasing diversity, governments might wish to take action to internalise the externality and encourage investment in non-fossil power plants.

Oil security externalities have been studied much more than security externalities in other fossil fuel markets. As noted above, they were found to be in the order of magnitude of 1-3 per cent of oil prices. In other words, they are fairly small. Gas and coal security externalities have received significantly less scrutiny. It is reasonable to assume that oil externalities are higher than those attached to gas and coal, since a large part of the externality is linked to market power in the world market. The size of any individual producer is small in comparison to total production, and market power does not appear to be a problem. The only cause of a major supply shortfall could possibly be an exceptionally long strike. Security externalities for coal can be assumed to be very small.

The damage caused to the economy by a supply disruption, and thus the size of the externality, is linked to both the extent and the duration of the disruption. Over the last half-century, the security of primary fuels was considered a serious problem only between 1973 and 1986. One way of addressing this energy security problem required investing in expensive plant capacity that lasts 30 to 40 years. If investment in long-term capital is used to

71. Environmental externalities would also have to be brought into the analysis. That is, the private cost of power generation from different fuels would have to cover the value of all externalities, including environmental externalities, in order to yield optimal fuel choice for society.

address much shorter-term supply disruptions, the cost of maintaining diversity could be much higher than the value of the externality it addresses.

Gas markets are regional, and although there are concerns about market power, these have generally been less than those prevailing in the oil market. The market power issues are most pronounced in Europe, which depends to a significant degree on a small number of outside suppliers, essentially Russia and Algeria with particular political problems[72]. A number of factors limit the impact of supply disruptions: gas can be substituted more easily than oil, and, outside North America, is traded under long-term contracts, which might help limit price increases. Most European countries have diversified gas supply sources where possible and can use interruptible contracts or draw on reserves.

Even under a worst case scenario, such as total disruption of Russian gas supplies to OECD Europe, which amounts to 25 per cent of the region's gas supplies, the impact of a year-long disruption on final supplies is small: the shortfall of gas to end users, on firm supply contracts, represents only 2.5 per cent of demand. The burden of this shortfall is unevenly distributed. Some countries are far more affected, but the result nevertheless shows that gas security externalities are either not very large or that they have already been partly internalised.

In most OECD countries, additional hydroelectric resources are limited. Nuclear power and non-hydro renewables are the main non-fossil generating options that could further diversify generation options. Most renewable energies either need no fuel supplied or the fuel is supplied domestically. In this sense, they offer security benefits. If either nuclear power or renewables could compete on electricity cost alone, they might be chosen without specific consideration of security externalities. In this case, they would automatically appear in the national fuel mix to help displace fuels that pose a security problem. However, neither appears to be

72. *For this and the following, see IEA (1995).*

the most economic generation choice in many countries. It thus makes sense to consider whether and where the improvement in fuel supply diversity due to their use might outweigh higher costs.

In the United States it appears that the cost gap between new nuclear and new gas plants is quite large: nuclear is at least two-thirds more expensive than gas-fired combined-cycle plants. Since the U.S. has comparatively large indigenous supplies of gas, one can assume that the security externality for gas is less than that quoted above for oil. This means that even taking into account security costs gas-fired plants would still be cheaper than nuclear plants and internalising the security externality would not alter investment decisions. In countries which are highly import-dependent, external security costs are higher, because the country is more exposed to market power in the international market. In such cases including security externalities might lead to different conclusions.

Conclusions: what Role for Governments?

Safeguarding energy supply security is an important element in any strategy designed to achieve sustainable development in the energy sector. The identification of an energy security externality – an impact of risk that is not adequately taken into account by private market participants themselves – clearly establishes a role for governments. This involves a short-run and a long-run component. In the short-run governments organise emergency responses to unexpected supply disruptions, such as the IEA's International Energy Programme. In the long run, governments can work directly on the political level in an effort to establish trust between producers and consumer and they can contribute through diversification and relative pricing to an energy mix that responds to the energy security externality. The means of intervention might be strategic stockpiling, import tariffs or the taxation of consumption.

This responsibility is not diminished by the difficulties involved in estimating precise values for security externalities. While it is necessary to be realistic about the level of precision that can be achieved in this sort of policymaking, the need for government action is well established. The tendency of liberalised markets to favour fossil fuels, especially gas, underscores the need for governments to remain vigilant in regarding adequate diversity in the energy mix.

Incorporating energy security externalities optimally into decision-making is further complicated by the need to internalise environmental and security externalities simultaneously. If externalities are to be addressed by governments, all of the relevant, identifiable externalities need to be internalised in a consistent, even-handed manner. This could alter the choices made in regard to diversity. On the other hand, simultaneously dealing with energy security and environmental externalities could offer synergies in the pursuit of sustainable development in the energy sector. For example, the reduction of gasoline consumption might well bring sustainability benefits in terms of both supply security and environmental performance.

The important point is not that one particular fuel or one particular instrument is necessarily superior to another. Different societies under different circumstances will make different decisions. Rather it is that maintaining energy supply security is an essential part of economic and social sustainability which needs to be fully integrated into energy policymaking.

ENERGY MARKET REFORM

Introduction

Structural and technological changes in the energy markets of IEA/OECD countries and beyond will have major, though not consistently predictable effects on sustainable development. Energy market liberalisation is widely expected to bring progress in the economic dimension, both through efficiency gains within existing systems and through enhanced technological dynamism. Its effects are less clear in the environmental and social dimensions. And if the issue of supply security discussed in the preceding chapter is to be adequately addressed by participants in liberalised markets, new policy instruments are likely to be necessary. In sum, the overall impact of energy market liberalisation on sustainability will be substantial, but it will be an evolutionary process that will unfold only in the long term.

This chapter lays out the principal axes of market reform currently transforming energy sectors in IEA/OECD countries, both generally and with particular reference to the electricity sector. It includes a definition of market reform in the present context, a discussion of its relationship to the three major dimensions of sustainable development, a brief historical overview of the power sector, and a discussion of key issues that continue to be of concern in regard to electricity market structure. Issues surrounding market reform in non-Member countries are discussed in Chapter 9.

Features of Energy Market Reform

What is Energy Market Reform?

The term "energy market reform" refers to the large number of processes underway in energy sectors all over the world that transfer the power of decision-making in energy industries from governments to private enterprises and consumers. Most importantly, it refers to the gradual substitution of more open and more competitive markets for publicly regulated monopolies. It also frequently includes the privatisation of government-held assets, such as controlling stakes in energy producing and distribution companies[73].

Because of their large size and part in the economic infrastructure, energy markets have historically been subject to substantial government involvement. Governments have been willing to take on greater project risk than private investors and they have sought to ensure efficient pricing by natural monopolies. These arguments have applied especially to electricity and gas markets: since disengaging from price control for petroleum products, governments have been less involved in oil and coal markets though some coal production continues to attract subsidies. Pursuing environmental objectives and supply security have been additional reasons for government involvement in all energy markets.

Technical and economic progress – most notably the fall in the unit size of power production equipment and the improved performance of information and communication technology (ICT) in metering and measurement – have facilitated the separation of production and distribution. This in turn has suggested new and more efficient forms of industrial organisation. However, the interplay between competitive production markets and monopolistic distribution markets remains a highly complex issue. The organisation of "power pools" in the electricity sector (see discussion below) and the question of "third-party access" in the gas sector are sensitive, hotly contested political issues involving billions of dollars and powerful stakeholders.

With growing numbers of players in energy markets, the need for government involvement in pricing has receded, though the need to

73. *IEA work on market reform has often concentrated on providing internationally comparable data and information on market reform to policy makers. See, for instance, IEA (2000),* Regulatory Reform: European Gas; *IEA (1999),* Electricity Market Reform: An IEA Handbook; *IEA (1999),* Electricity Reform: Power Generation Costs and Investment; *IEA (1999),* Regulatory Reform in Argentina's Natural Gas Sector; *IEA (1998),* Natural Gas Pricing in Competitive Markets; *all OECD/IEA, Paris. In a second phase, analysis of market reform processes will explicitly address sustainable development issues. See background paper for the IEA conference "Energy market reform and the environment", 19 February 2001, Paris (forthcoming).*

internalise energy externalities remains a function of governments. The growing size and sophistication of financial markets also enable private investors to deal more easily with commercial project risk, though the management of supply risk remains an ultimate responsibility and a political problem, for governments. Competition also encourages the creation of new products, such as "green electricity".

How Does Market Reform Affect the Three Dimensions of Sustainable Sevelopment?

Market reform, which is characterised by increasing emphasis on profit maximisation and the search for greater efficiency by producers, can have profound impacts on the magnitude and structure of energy supply. This in turn has important implications for sustainability in the energy sector as different energy supply structures imply different impacts on the environment and on the security of energy supply.

The effects of market reform are felt mainly through four channels:

- the absolute magnitude of energy demand;
- the technical efficiency of energy production and transformation;
- the fuel mix in primary energy supply and in power generation; and
- the creation of new energy markets and services.

In general, it is believed that market reform and deregulation will reduce energy prices, particularly electricity prices, through increased competition. Lower energy prices would translate into gains for energy consumers and improved economic growth. In many liberalised markets, however, these gains have been relatively modest. For example, an estimate for the United States, which has extensively liberalised power markets, indicates a two per cent decline in prices over the last 10 years. However strong the price impact is in practice, it will, other things being equal, contribute to an increase in energy consumption and, in the absence of changes in technology and the fuel mix, lead to increased environmental impacts. In addition, the impact on supply security could be negative, particularly in the European gas market, with its growing dependency on a small number of suppliers (see graph below)[74].

Figure 11

OECD Europe Net Imports of Gas by Origin

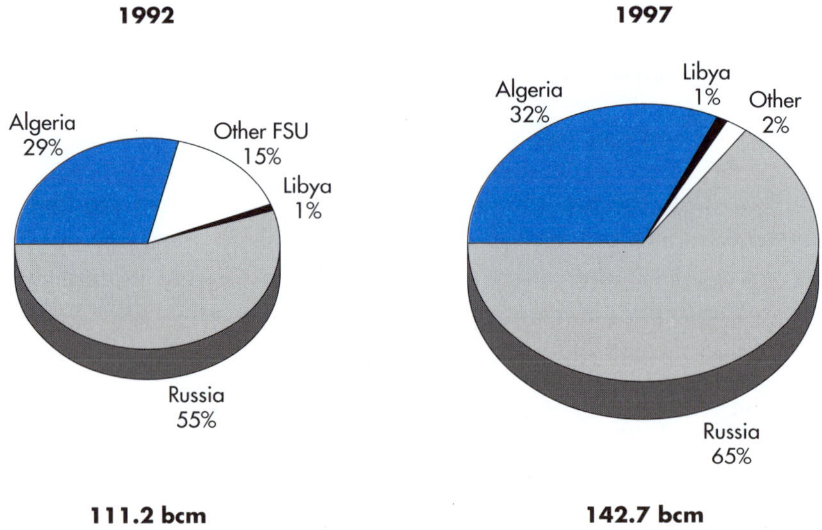

Accelerated technological progress due to an infusion of entrepreneurial dynamism and increased competition offers much hope for a positive contribution of market reform to sustainable development. Two factors might limit such a contribution: (1) the "rebound effect" could lead to further demand increases if the effective price of energy services falls; and (2) the need to "squeeze" assets might induce competitors to retain existing equipment even if better performing technologies are already available.

Nevertheless with the right incentives in place competitive enterprises can quickly take advantage of technical and structural

74. *IEA (2000),* World Energy Outlook 2000, *p. 148. See also the information on oil supply security provided in Chapters 2 and 3.*

changes that deliver environmental benefits. A case in point is the U.S. sulphur dioxide trading programme, which has delivered very substantial emissions reductions at much lower than estimated costs due to price declines in scrubbers and the availability of low-sulphur coal.

Market reform can also dramatically affect the fuel mix. This is particularly visible in the European power generation market, where it is predicted that the political acceptability and the improved economy of combined cycle gas turbines (CCGTs) will allow gas to make large in-roads at the expense of both nuclear power and coal (see graph below)[75].

However, the fuel mix in a liberalised market will in the long run also be affected by non-energy developments, especially in power generation. Interest rates can play a decisive role in determining the speed and the structure of investments in new capacity and the retirement of old ones. Generally speaking, higher interest rates favour existing installations even if they are inefficient over new construction, and new investments with low capital costs over those with high capital costs. In most cases, gas-based generation has the lowest capital costs relative to fuel prices, followed by coal. For nuclear power and renewable energy, instead, fuel prices are almost negligible[76]. High interest rates thus favour gas, while renewable energy sources, in particular, would profit from lower interest rates. In formerly protected markets, decision-makers were largely sheltered from such considerations, as any changes in relative input prices could be rolled over to customers. This is no longer the case and economic considerations will accelerate structural change over the coming years.

Finally, market reform encourages satisfying consumer demand with new products and services. Two kinds of potential new markets are taking shape:

---

75. Ibid., p. 140.

76. *This consideration abstracts from maintenance costs. Including them, however, would not change any of the qualitative relationships.*

Figure 12

Total Primary Energy Supply in OECD Europe by Fuel

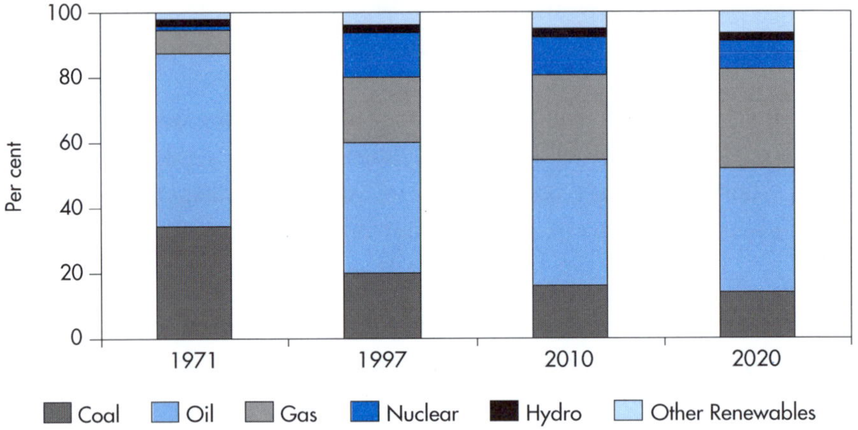

- Energy service markets, in which customers will contract with providers not over amounts of energy delivered, but over performance in terms of energy services mainly lighting and heating. Demand-side measures, such as more energy-efficient lighting, are also possible. Such energy service markets could provide further economic efficiency gains that could contribute to the environmental and social dimensions of sustainable development.
- Markets for particular forms of energy. "Green electricity" or "reformulated gasoline" are examples of energy forms with reduced environmental impacts. In some liberalised power markets the branding of electricity has led to a large array of electricity choices, most of them competing over environmental criteria and various pricing formulas (see box below). Such markets allow consumer preferences for improved environmental performance to be taken into account.

The growth of these kinds of markets offers new opportunities for the environmental dimension of sustainable development, but this type of approach alone will not solve environmental issues. First, the environment remains very much a "public good" issue. This means consumers will buy "clean energy" up to the point where the price they pay matches their own "feel-good factor," and not up to the point where their efforts take account of social benefits to everybody else. Second, there are currently no clear standards to allow truly informed consumer choice; claims about "green electricity" reflect marketing gimmicks as much as true product differentiation. For both reasons, governments have an important role: first, continuing to perform their traditional task of internalising externalities, which can now be aided by individually motivated buying decisions; and second, setting clear and transparent performance standards that will allow consumer comparisons and quality competition.

Box 8

Product Differentiation in Deregulated Electricity Markets

Deregulation in the electricity sector, unbundling and new entrants have introduced new forms of competition at the retail level. These include not only price competition, but also quality competition. While one kilowatt-hour is physically indistinguishable from another, the basis on which it is generated (fossil fuels, nuclear, renewables...) can vary enormously and offers great scope for product differentiation. The main criterion for product differentiation in power generation tends to be environmental performance.

The Internet Website of "Green-e" at www.green-e.org/power/cert.html, an information platform for power providers, offers, for example, more than a dozen different products for retail consumption. They range from Greensmart *(100 per cent biomass and geothermal), and* Enron Earth Smart 50 *(50 per cent renewables, 50 per cent non-coal, non-nuclear) to* EnviroBlend *(5 per cent New Renewables, 45 per cent Renewables, 50 per cent Hydroelectric) and* GreenValue 100 *(100 per cent Renewable Energy from Sun, Earth and Wind). In Germany more than 20 national providers compete, many of them using renewable energy sources. This sort of product innovation presents an opportunity for sustainable development, by enabling consumers to exercise choice between products with different sustainability performance.*

A Short History of Electricity Market Reform [77]

Energy market reform is a relatively recent phenomenon. It is instructive to review briefly its history in the electricity market – the most dramatic example of energy market reform – in the context of sustainable development. The review demonstrates that even in markets that are national due to their physical nature, there exist broad trends across countries.

The origins of the current wave of reforms in the electricity sector go back to the late 1970s. The first step was a partial opening of electricity generation to new entrants. In 1978, the U.S. adopted the Public Utility Regulatory Policies Act (PURPA), requiring utilities to buy electricity from "qualified facilities", mostly co-generators and small power producers. Four years after PURPA was passed, Chile enacted a law introducing some competition into electricity markets by allowing large end-users to choose their supplier and freely negotiate prices.

A second step came with the establishment of explicit market mechanisms to determine the dispatch of generators and the wholesale price of electricity, thereby permitting competition between generating companies. The England and Wales electricity market, or "pool", established in 1990, was the first such mechanism. In the following years several other OECD countries followed.

Competitive power exchanges started operation in 1998 in Spain and, within the U.S., in California and through the "Pennsylvania-New Jersey-Maryland Interconnection". This opened the way for a number of regional electricity markets within the U.S. and Europe. In parallel to the development of wholesale markets, electricity markets have been progressively opened up to end-users. In some countries, end users are legally permitted to choose their supplier. In others, there is some degree of market opening, even if no organised electricity market has been established.

77. *The material in this sub-chapter is taken from* Competition in Electricity Markets: An IEA Handbook, *p. 19.*

Key Issues in Market Reform

As governments plan for the introduction of competition, in particular in electricity markets, they are confronted with a large array of options. One expert counted 15 basic market structures, with up to 30 variations in their details. A primary consideration in establishing a competitive electricity market is the number of potential competitors and how to ensure that they engage in true competition.

A basic choice in introducing competition is whether a mandatory competitive pool has to be established, such as in the UK, or whether market participants should have the possibility to engage in physical bilateral power exchanges outside of the pool, such as in the new Spanish system. Under certain conditions, these two systems should lead to the same, efficient, result. These conditions are good access to information for all market participants, an independent system operator for the integrated transmission area, and sufficient competitive pressure on all players in the market. The co-existence of a pool system and bilateral contracts could bring benefits in increased competition, as well as wider choice for generators and consumers.

Competition in generation is a key aspect of electricity reform. Under competition, generators typically have the option of entering their supply prices into a general pool of electricity in which a dispatch merit order, based on the bids that have been received, is established. Electricity pools are now in operation in England and Wales, Norway, Australia, Spain, Canada (Alberta), Chile and Argentina, among others. Electricity pools can be mandatory as in England and Wales, or non-mandatory as in Norway, in which case bilateral trade outside the pool is permitted.

The physical nature of electricity does not allow a true electricity spot market, i.e., a market for immediate electricity delivery. Pools are a substitute for true spot markets. Pool purchasing prices and supply schedules are set by auction some time in advance of physical delivery (e.g., one day or one hour in advance). Pool selling prices are established by adding the cost of imbalances, ancillary services, and possibly other demand-related charges (e.g., peak demand availability payments) to the pool purchasing-price. Since prices are determined by scheduled supply and demand, these are known as *ex ante* pools. Alternatively, in an *ex post* pool (e.g., as in Australia) prices are determined by actual generator schedules and demand.

Electricity pools perform two different functions. First, the pool manages the technical operation of the system such as dispatch and ancillary services. System operation is centralised in virtually all electricity systems due to technical constraints. Decentralisation would impose very high costs of bilateral co-ordination. Second, the pool determines an economic merit order among pool generators and a price. In principle, a System Operator and a Market Operator (or Power Exchange) can separately perform these two functions; efficient operation would, however, require substantial co-ordination between them.

Pool performance critically depends on the horizontal and vertical structure of the market. A number of comparable generators is necessary to induce competitive bidding behaviour. Vertical integration of generation and transmission assets instead may hinder competition, as a transmission network operator may have an incentive to favour its own generation assets. Pool performance is also affected by the degree to which generators are allowed to enter into long-term contractual relationships to manage their risks. Most contracts take the form of forward contracts or Power Purchase Agreements (PPAs). Contract markets may change the incentives of generators to exercise market power in the pool. PPAs for instance may change competitive behaviour due to the forward integration of generators and retailers.

Pool regulation strives to achieve two complementary goals. First, it must provide open and non-discriminatory access to the pool for all players. Second, it must allow efficient system operation in both the short and the long term. Ensuring adequate access requires that system and market operation be independent from market players, whereas efficient system operation requires a relatively large degree of co-ordination between pool operation and pool players. The challenge for regulation is to maximise co-ordination while preserving pool independence.

System operation has one prominent feature from the point of view of regulation: it is a centralised (i.e., monopolistic) activity. This network nature of electricity distribution (and, equivalently of gas distribution) continues to require a regulatory role of government. To prevent monopolistic behaviour, the ownership, structure and operation of the system operator need to be regulated. In particular, regulation needs to ensure that the system operator does not exercise market power in order to generate excess profits, nor discriminates among market players. In general, this includes barring the system operator from engaging in the production of energy for its own account[78].

Concluding Comments – Sustainability Policies and Market Liberalisation

The impetus for change and the key issues driving energy market liberalisation are primarily contained within the economic dimension of sustainable development but there is scope in the process of market liberalisation for synergies or win-win options between the three dimensions of sustainable development. Privately profitable technological progress, such as in the area of energy efficiency, might well also further public objectives such as the reduction of greenhouse gases. Imports of electricity generated from renewable energy sources, like Norwegian hydropower, may substitute for more expensive and environmentally less benign sources like Danish coal-fired power plants. Competitive markets can facilitate solutions in the environmental and social dimensions of sustainable development, though they do

78. For further information about the structure and management of power pools see Von der Fehr, N. and D. Harbord (1997), "Competition in Electricity Spot Markets: Economic Theory and International Experience"; Hunt, S. and G. Shuttleworth (1996), "Competition and Choice in Electricity"; and Joskow, P. (1997), "Restructuring, Competition and Regulatory Reform in the U.S. Electricity Sector".

not necessarily do so. In general, the pursuit of social and environmental objectives in relation to market activities continues to require the involvement of governments. That is in part why liberalisation generally does not involve complete deregulation.

It is obviously important, however, for governments to consider new ways beyond direct regulation to satisfy environmental and social objectives related to energy markets. The obvious option is to shift to greater use of market-based policy instruments, such as taxes, subsidy reform and emissions trading schemes. They can be less costly and more effective in many situations. But there are opportunities to do more. Important in this context is the discussion in Chapter 2 on the need for a dynamic approach to dealing with externalities. Internalisation of externalities can take place through structural adaptation, as well as through the classic Pigovian tax-based model. In this framework, market reform can create new opportunities to deal with environmental and even social issues in creative ways. Legal structures can be altered, new technologies can be applied, public information systems can be enhanced – a variety of new ways can be designed to link environmental and social objectives to new market structures. While market liberalisation in one sense reduces the role of government in the energy sector, in another it offers a new positive opportunity for policymakers.

IMPROVING ENERGY EFFICIENCY

Energy Efficiency and Sustainable Development

Being efficient in the use of all resources makes an important contribution to both environmental and economic sustainability. Attention was sharply focused on the importance of using energy efficiently during the first oil price shock in the 1970s. Since then, total final energy consumption per unit of GDP in IEA countries has fallen sharply. Today IEA economies use about 45 per cent less energy than in 1973 to generate one unit of GDP. This decrease has been driven by improved energy efficiency in key end-uses and by shifts in economic structure and consumer behaviour.

A variety of mechanisms is available for supporting the development and adoption of new energy-efficient technologies. These include direct funding for research and development, the setting of industry standards, changing relative prices through subsidies or taxes, labelling programmes and the branding of efficiently produced energy services in order to allow informed consumer choice. This chapter reports on progress that has been made in improving energy efficiency and provides an overview of the policy experiences so far. It ends with a list of elements that are important for a successful policy strategy in the area of energy efficiency.

Progress in Energy Efficiency

Higher fuel prices, long-term technological progress and the energy efficiency programmes of governments have been important factors in bringing about improvements in energy efficiency. Energy use per unit of GDP (overall energy intensity) fell most rapidly during the first years after the 1973 oil shock, a result of higher energy prices and more widespread implementation of governmental energy efficiency programmes. Oil prices peaked in 1982 and remained high until the reverse price shock in 1986. Surprisingly, overall energy intensity fell less during the years 1982-86 when prices were high than in the four following years, averaging 2 per cent and 2.3 per cent per year respectively (see Figure 13). In European IEA Member countries this trend was even stronger – overall energy intensity fell by only 0.5 per cent per year on average between 1982 and 1986 and then by 2.3 per cent per year through 1990. After 1990, the annual decline in energy per GDP slowed considerably, but only until 1996. Between 1996 and 1998, overall energy intensity in IEA countries fell by more than 2.5 per cent per year on average.

The fact that the fall in energy prices after 1986 did not lead to an immediate slowdown of the decline in overall energy intensity may suggest that much of the efficiency gain up to 1986 was of a permanent, technological nature. These new technologies continued to work their way into production processes and consumer technologies afterwards even though average prices had fallen. It might also suggest a time lag in the normal process of adjusting to changes in relative prices. Only careful and detailed research of the actual decision-making processes can indicate which hypothesis, or which combination of the two, provides a better guide.

Changes in economic structure, such as lowering the share of GDP contributed by manufacturing or moving the mix of manufacturing output away from energy-intensive products, affect the energy-to-GDP ratio. This was the case during the 1970s, when the output share from energy-intensive heavy manufacturing declined in some IEA economies. Since 1982 this trend has been less prominent, with the output from both total manufacturing and heavy industries roughly following overall GDP trends. This suggests that, for IEA countries taken together, changes in economic structure played a relatively less important role in changing overall energy intensity after oil prices peaked in 1982 than before.

Many changes in consumer behaviour affect energy use and its relation to GDP. Factors like increases in the ownership and use of

Figure 13

Total Final Consumption per GDP for IEA-Total and IEA-Europe (1982 = 100%)

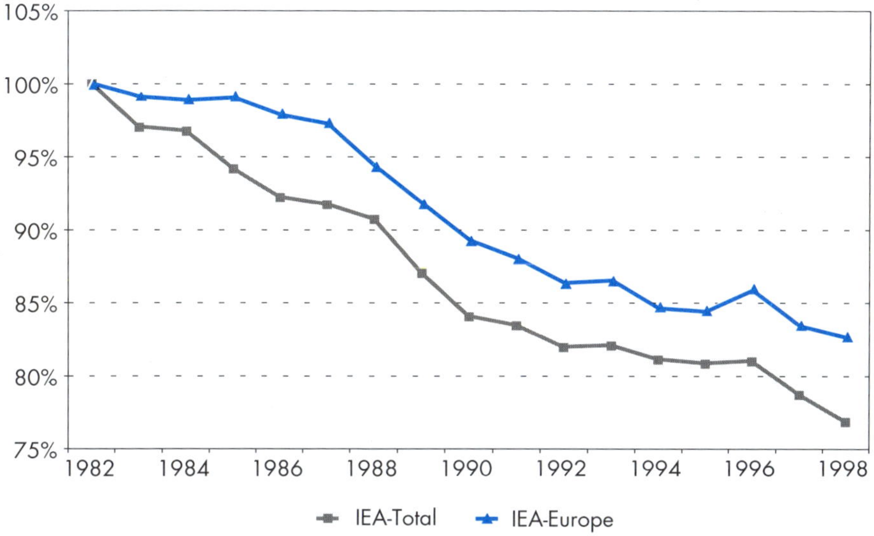

electric appliances, bigger houses and more personal travel by car drive energy use and affect overall energy intensity. For example, in the 1970s and 1980s the number of person-kilometres of travel grew more than GDP in many IEA countries and translated into increases in energy use per unit of GDP.

Using more disaggregated measures than energy per unit of GDP allows the analyst to unravel most of the effects of changes in economic structure and behaviour and thus provides better estimates of basic energy efficiency developments. Investigating energy intensities at the levels of sub-sectors or end-use sectors – such as the amount of energy needed to produce a tonne of steel, to heat a square metre of floor space or to travel a kilometre in a car – shows that these have declined significantly over the last 25 years. Although in most countries the strongest falls in these intensities occurred during the 1970s, intensities have continued to decline in many countries despite the lower energy prices after 1986.

The ratio of total primary energy supply (TPES) to economic output has fallen less than total final energy consumption (TFC) because of the increasing use of electricity. In some cases, energy losses in electricity generation outweighed reductions in energy intensities that occurred when direct fuel-fired applications were replaced by electric end-uses having higher end-use efficiency. Most of the increase in electricity use is, however, driven by the use of more electricity-based services and appliances. The increased use of electricity means that a higher share of the energy supplied today is consumed in the electricity sector. As electricity demand is expected to continue outpacing total demand, improving the efficiency of both electricity supply and electricity end-use is ever more important.

The share of oil in the energy supply of IEA countries has fallen significantly, from 55 per cent in 1973 to 41 per cent in 1997. The decline in oil use is evident for all stationary purposes, most notably in electricity generation, where oil in 1997 only represented 6 per cent of IEA generation, down from 25 per cent in 1973. Today oil products constitute about 30 per cent of industrial energy demand and about 20 per cent of demand in the household and services sectors. Transport relies almost entirely on oil and accounts for an increasing share of total oil demand. Hence energy efficiency policies in the transport sector will be the most important in regard to reducing oil dependence. Regrettably they are also the most difficult to implement. (This is discussed further in Chapter 7.)

Progress in Residential and Commercial Buildings

Residential and commercial buildings account for about one-third of total final energy use in IEA countries. While the share of buildings in IEA oil demand is relatively low (11 per cent), building electricity consumption accounts for almost 60 per cent of total IEA electricity demand. Space heating is the most important energy end-use in residential buildings for most IEA countries. The significance of space heating varies according to climate. Figure 14 shows *per capita* residential energy use with space heating normalised to a similar climate for selected IEA countries. The amount of energy used for space heating differs widely among countries due to differences in factors like house size, indoor heating comfort, heating equipment used and insulation levels. The fastest growing end-use in buildings is electric appliances. The growth of this end-use is expected to continue and will put pressure on electricity supply. Water heating also constitutes an important part of residential demand, with its fuel choice often linked to that of the space-heating system.

Figure 14

Residential Energy Use per capita (Climate adjusted)

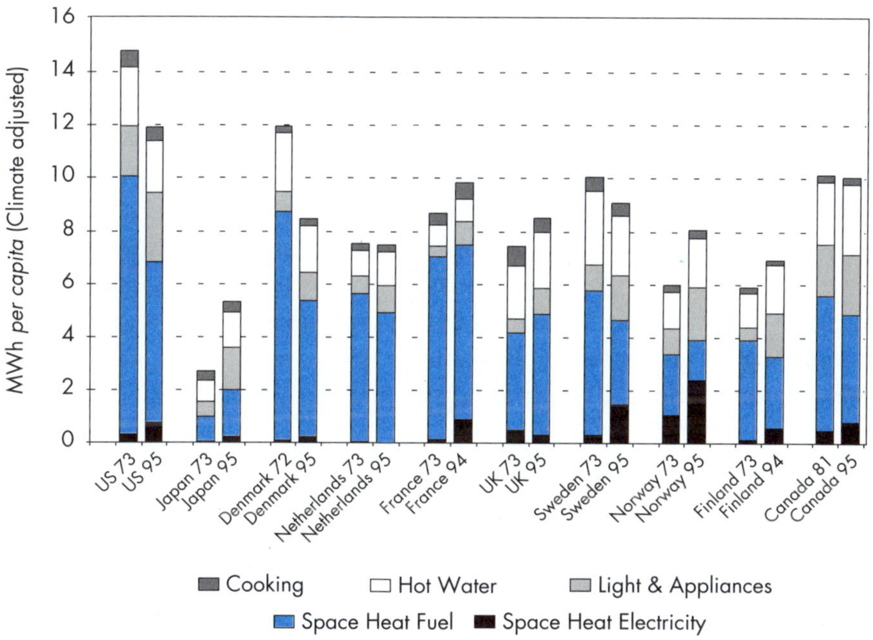

Most IEA Member countries have promulgated policies to reduce energy use for heating. A number of countries have instituted regular, detailed household heating surveys to follow the progress of heat energy-saving efforts. Figure 15 portrays a key indicator of space heating that captures the changes in heating, adjusted to take into account the approximate efficiencies of combustion equipment. The intensity shown is normalised to house area and to climate variations using degree-days, to make the heating figures comparable both year-to-year within a country and among countries. The data in the figure suggest that many countries have made significant progress toward improving space-heating efficiency. In the U.S. and the Netherlands, for example, savings of about 25 per cent were achieved between 1982 and 1995 due to reductions in space heating intensity. In some cases the reductions shown in the figure may actually underestimate savings as many countries experienced increases in the standard of indoor comfort through heating larger parts of buildings, heating longer hours and to higher indoor temperatures.

Figure 15

Space Heating Intensity
(Useful energy per square metre, per degree-day)

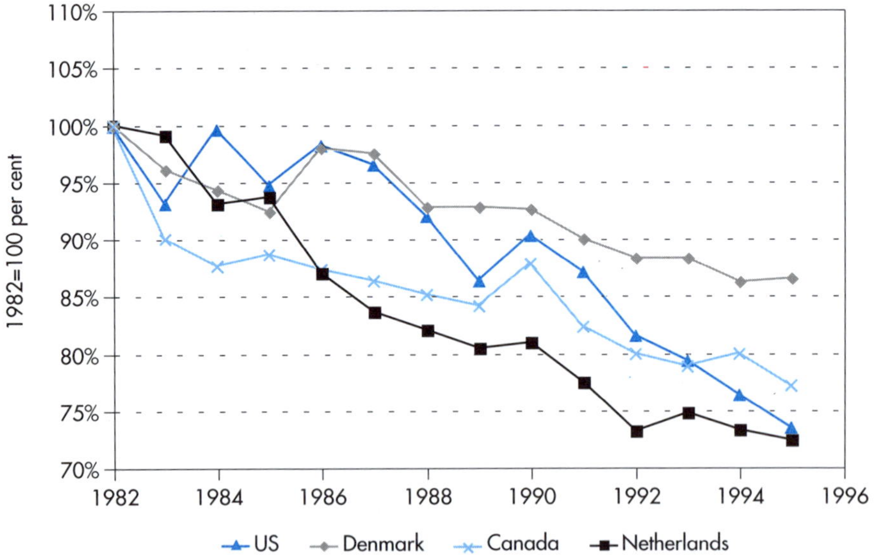

Energy Efficiency Progress in Manufacturing

In the aggregate, energy use relative to manufacturing output (measured by value added) has fallen more or less continuously in most IEA countries since the 1950s. This intensity in the aggregate data is affected by structural shifts toward or away from energy-intensive products and by changes in individual energy intensities in each manufacturing sub-sector. Governments have encouraged the latter effect.

Figure 16 shows how manufacturing energy use per unit of value added in selected IEA countries, with impacts from structural changes accounted for, has evolved between 1982 and 1995. All countries represented in the figure achieved significant energy savings due to reductions in energy intensities. In all cases, energy

Figure 16

Manufacturing Energy Intensity (Corrected for structural changes)

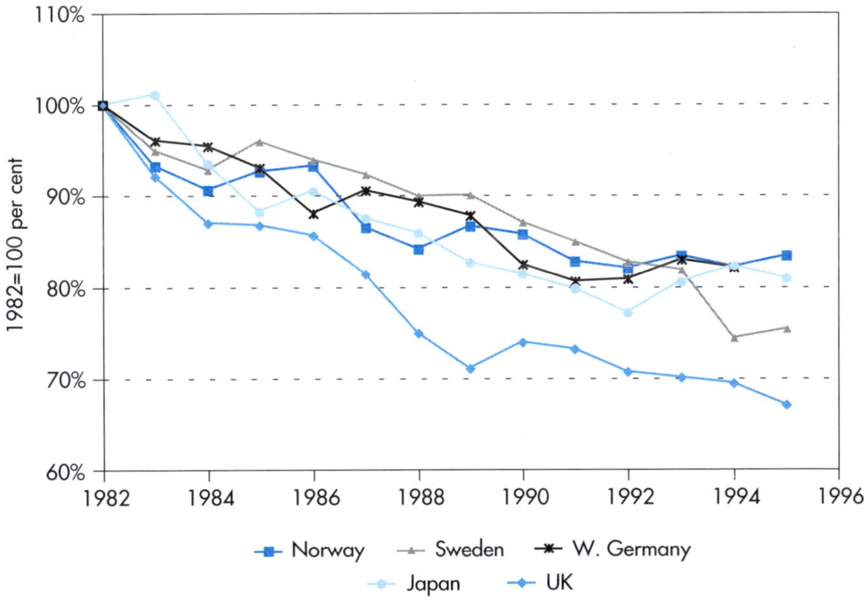

intensities continued to decline after oil prices collapsed in 1986. In fact, long-term data show that structure-adjusted intensities also fell before the rise in world oil prices in the 1970s. These savings were largely the result of improvements in technology and increases in the scale of manufacturing of various products.

The reductions in energy use per value added can be confirmed by investigating intensities expressing energy use per ton of physical product. For example, the energy requirements to produce one ton of steel have fallen significantly due to improvements in steel-making processes. Comparing absolute values of these intensities among countries and/or companies may reveal considerable differences in practices that can save energy, of which the most effective are often called "best practices". While the technologies behind best practices are important to understand, it should also be borne in mind that "best" is not only a function of energy intensities but also the cost of other resources, like capital and labour.

Energy Savings Potential

While energy efficiency is widely viewed as an important element of energy and environmental policy, there is little agreement on specific energy efficiency goals and the best ways to achieve them. The lack of consensus stems in part from differing views about the meaning of energy efficiency.

Energy efficiency refers to the relationship between the output (service) of a device or a system and the energy put into it. For a motor, it is expressed as the percentage of input energy converted into useful power output. For an automobile, it is often expressed in litres of fuel per 100 km. This is only one of many output indicators; others include the number of passengers and the weight of the cargo carried. Improved energy efficiency is doing more with equal or less energy input, e.g., more goods produced, fewer kilowatts per tonne of aluminium used, more travel, comfort, light provided.

Estimates of energy efficiency improvements and their impact on energy demand are based on assumptions about technical factors, equipment costs, expected rates of market penetration, consumer behaviour and policy measures. When considering potential for improvement, it is essential to distinguish between the realm of technological achievements and the real world of consumers.

Definitions of cost-effective or economic potential usually assume an ideal world where producers and consumers act in an economically rational way and adopt energy-efficient technology as soon as it becomes "cost-effective". Many studies base their estimates of energy saving investments on a comparison between the cost of energy saved and that of energy produced, using the same discount rate. Where the cost of conserved energy is lower than the supply costs, the energy efficiency investment is considered cost-effective and it is assumed that the saving will be made by the consumer or even by the producer.

The evaluation of the cost-effectiveness of an energy efficiency improvement depends largely on the discount rate used, though there is no agreement on what represents an appropriate discount rate, as discussed in Chapter 2. In some cases the discount rates applied to energy efficiency improvements are the same as those used by utilities for energy supply investments; in other cases, premiums are included to take into account resource depletion, energy security and environmental considerations. Most individual consumers make investment decisions without direct reference to discount rates and discounted capital flows. Even in business or industry, where investments are more likely to be evaluated in terms of rates of return or payback time, energy users may apply more stringent investment criteria to energy efficiency investments than to productive investments and those that increase their market share.

Even when the apparent costs of energy efficiency are much less than those of new energy supply, investments in efficiency are often more difficult to finance. Suppliers and users of energy are two different groups with vastly different investment priorities and access to capital. Many efficiency measures that would pay for themselves in two years or less do not appear financially beneficial to the energy users who ultimately make the investment decisions.

As energy efficiency is often a minor consideration in the choice of energy-using equipment, its costs have to be paid back at a rapid rate. Product characteristics other than energy efficiency are usually more important for individual consumers. In addition, investments in energy efficiency are subject to fluctuations of energy prices. Information on the performance of energy efficiency investments is often difficult to acquire. They are perceived as having a higher risk than many other business operations. These issues contribute to the problem of identifying a single, absolute or cost-effective potential.

Buyers do not normally base their choices on formal economic calculations only, but also on considerations of comfort, quality and availability of a product. Nevertheless a calculation of cost-effectiveness gives valuable information about how well resources are used and if their use can be improved. The cost-effectiveness of an efficiency measure depends on which costs and benefits are considered. For example, from a business perspective, the only relevant costs and benefits are those borne by the energy user. These include the expenditures for equipment, engineering and installation as well as charges for production downtime. The benefits include energy cost savings, plus enhanced labour productivity, environmental compliance or product quality. These are the traditional accounting costs and benefits that directly affect the firm's bottom line.

From a societal viewpoint, there is a wider range of relevant costs and benefits. These include monetary, health, and ecological costs and benefits that accrue to society. Certain societal benefits, such as reduced local air pollution or diminished global warming are external to the market and are difficult to quantify. Moreover, they accrue to society at large, not to the particular party implementing the efficiency measure. This wider definition of cost-effectiveness is the more important measure for policy aimed at energy security and environmental quality.

Taking account of cost-effectiveness, energy cost savings, higher purchase prices, and sometimes other less explicit cost considerations, a framework can be provided for describing the potential energy savings and CO_2 reductions due to improvements in energy efficiency. Different estimates of potential are:

- *Market potential* is the saving that can be expected to be realised in normal practice. It reflects what is seen to be technically and financially viable by individual decision-makers. (Note that in many energy demand and CO_2 emissions forecasts this potential is often included in the "baseline".)
- *Economic potential* is the saving that can be achieved by optimising costs relative to best available practice. It reflects the viewpoint of individuals and organisations.
- *Social potential* is the saving that might or might not be profitable from a private point of view but that would yield a positive net benefit for society as a whole. In this case externalities are taken account of. An example would be an energy saving that was facilitated by a subsidy to consumers with a view to reducing greenhouse gas emissions.
- *Technical potential* is the achievable savings resulting from the maximum energy efficiency improvement available at a given time, regardless of cost considerations.

These definitions point to the importance of market actors in determining potential energy savings. There are many things that individuals or organisations can do to become more energy efficient. There are still more options, involving trade-offs among multiple economic entities, available at the level of the whole society.

Nearly all energy-using devices and systems are less energy efficient than their theoretical maximum efficiency. There is always potential for improvement. A device or system can only be more or less energy efficient than the alternatives available at a given time in a given situation. But conditions change – innovation produces improved devices and systems; deployment patterns shift and result in different economies of scale; energy sources and prices change; government policies and programmes change. Today's efficient device is rarely tomorrow's.

Energy Prices

The role of the price of energy is fundamental. It is a major criterion on which consumers assess whether energy-saving measures are worthwhile. Another factor is that investing in efficient energy end-use technology can require major capital expenditures; efficient systems and equipment are usually, though not always, more expensive than the technology they replace. Among the many comparisons consumers have to consider when buying appliances or equipment is whether or not the energy cost savings of more efficient models are worth the higher purchase price of those models. Obviously, the higher the price of energy, the more attractive is the investment in the more efficient model.

The more energy prices reflect the full costs of producing energy and mitigating the environmental damage it incurs, the more potential there is for energy savings. This is illustrated in Figure 17.

Figure 17

Energy Efficiency Potential Related to Price

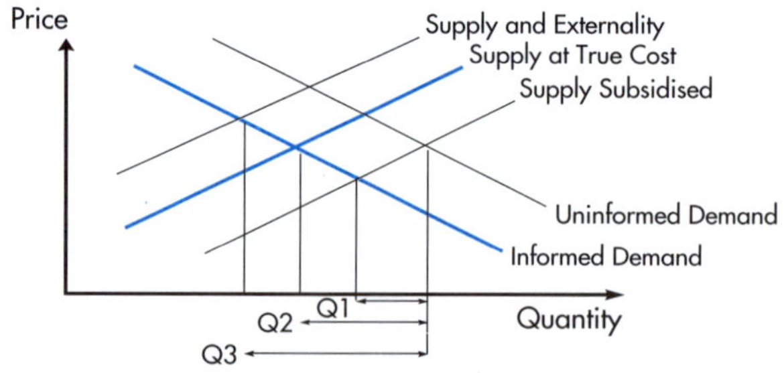

Note: The more energy prices reflect the full costs of producing energy as well as the environmental damage that energy generates (externalities), the more potential there is for energy savings. However, when prices are distorted by subsidies, energy consumption will not reflect its true cost to society and the potential for energy efficiency improvements is not sufficiently exploited. See IEA (1998), Energy Efficiency Initiative, *p. 86.*

Efficiency Gap and Market Hurdles

Despite all potential for energy savings, there is still a phenomenon known as the "payback gap". For various reasons, cost-effective energy savings options are often not taken. This under-investment in energy efficiency suggests that many factors other than direct quantifiable costs (purchase prices, maintenance costs, energy costs) affect consumer decisions.

This gap exists for all types of energy consumers. Individuals and organisations invest less in energy efficiency than would seem economic because of lack of information and awareness, lack of technical personnel, lack of investment funds, uncertainty about energy prices and uncertainty about equipment performance. They are also discouraged by equipment supply infrastructure problems, and the split incentives problem, also know as the tenant-landlord problem. This occurs when one party pays for the equipment (commonly the landlord) while the other party pays the energy bill.

A similar problem occurs in regard to investment in electricity generation plants when lower overall costs may be achieved through energy efficiency investments. A power system can meet new demand by adding more supply capacity or by reducing surplus demand. From an overall societal point of view, good economics requires that the least-cost option should be chosen. The investment dilemma here is that different actors make the choices: on the supply side, it is the company and on the demand side it is the consumer. They have different perspectives on their choices. The suppliers have more opportunity to obtain and treat information and to spread their risks than do consumers. Supply of energy is their core business. In their calculation they know that their installations will be used for a long time and have a high degree of flexibility to meet different demand patterns. In some cases, they even have a dominant position in the market or a monopoly.

Individual consumers often have less information and fewer opportunities to act optimally over the long run. Energy use is just one component of the desired service. Investments may be made at lowest first-cost without due consideration to lifecycle costs. The difference between the two perspectives is called "the pay-back gap".

This "gap" suggests a role for policy measures focused on stimulating innovation and investments in energy efficiency improvement. Such policies would include accelerating technology development and demonstration, stimulating product demand through procurement activities, applying efficiency standards to information-poor sectors, influencing consumers' energy-using behaviour, and in general finding ways to create markets for energy savings, thus stimulating further innovation.

The reasons for the payback gap are debated among energy efficiency professionals. The sometimes disappointing levels of adoption of energy efficiency technologies have been analysed and some progress has been made on understanding the considerations that underpin consumer resistance.

Infrastructure, Capital Stock Turnover and the Potential for Change

Existing energy and transportation infrastructure, slow capital stock turnover and a general resistance to change lock in energy use patterns for long periods. Standardised patterns of consumer behaviour may also slow the adoption of energy efficiency improvements, even where costs of energy efficient products are becoming more attractive, as in the case of compact fluorescent light bulbs (CFL). A 15-watt fluorescent lamp provides the same amount of light as one 60 watt incandescent lamp but uses only 20 per cent of the energy and lasts eight times as long. When the price of the lamp is considered along with the costs of electricity and maintenance, it can be shown that the CFL is a cheaper alternative. Moreover, the cost for CFLs today is lower than ten years ago. This is because volume has grown, which incites better technical solutions with lower per-unit-costs for manufacturing through the exploitation of dynamic economies of scale. This phenomenon is referred to as the learning curve (see Chapter 6 for a more extensive discussion of learning curves)[79].

---

79. *IEA (2000h)*, Experience Curves for Energy Technology Policy.

This learning effect can be exploited systematically with use of energy policy programmes for research and development, demonstration, technology procurement, and labelling. A projected growth in sales volume will have positive effects on the costs and the prices. An accelerated growth in volume will speed up these effects. Such opportunities occur when procurement is made by central purchasers or by a group of co-operating local purchasers.

Some observers express great hopes for a quicker uptake of new energy- and cost-saving technologies due to the competitive pressures associated with energy market reform. Similarly new products and services may be created by energy service companies. It will still take a few years before it can be clearly seen how energy efficiency and investment patterns will stabilise.

Toward an Effective Energy Efficiency Policy

Those who work on energy efficiency analysis in the IEA have identified a series of elements that should be part of strategies intended to promote improvements in energy efficiency.

Establish and Maintain an Effective Market Structure

The majority of energy efficiency decisions are made in market transactions. It is the millions of consumers, producers and market intermediaries who ultimately decide through their behaviour, purchases, product designs, research activities and investment decisions how efficiently energy is used. The potential for energy efficiency improvements can never be fully realised unless these energy users take into account, either explicitly or implicitly, the energy and economic consequences of their everyday decisions. Markets can also provide a useful framework for weighing the importance of energy efficiency against other individual and social concerns. In some situations, transition to full cost pricing and the removal of subsidies may cause social difficulties. In these cases, transitional policies, such as "life-line rates" for essential household consumption, may be necessary to give market actors time to adjust. Actions include:

- establishing real cost pricing;
- removing subsidies and cross-subsidies;
- using taxes and levies; and
- securing clear rights and responsibilities of property ownership.

Help Market Actors Recognise their Best Interest and Act on it

Evidence shows that energy users adopt fewer energy efficient options than are in their financial interest. Part of the reason is that consumers often lack the information, time and skills to recognise their own interest in energy efficiency. After receiving proper economic signals from the market, consumers need information and expertise to make better informed, energy efficient purchases and behavioural choices. Other market actors may also need help in acting on their interests in energy efficiency. For example, in a true market, energy service companies (ESCOs) or other intermediaries may be willing to tap some energy savings opportunities, but may need help in the form of technical training and general business advice. Actions include:

- analysing consumer behaviour;
- providing information and/or training on the energy efficiency aspects of products and behaviour;
- encouraging the formation of energy service companies and third-party financing; and
- creating special financing arrangements.

Focus Market Interest on Energy Efficiency

Though market forces and information are very important in increasing the rate of energy efficiency improvements, there are market failures and barriers that can inhibit efficiency gains. For example, information may be difficult to convey to end-users, especially if the purchasers are not the users. Or consumers may not have adequate market power to show manufacturers their interest in the energy efficiency features of products. In such cases, certain government interventions, such as codes, standards, voluntary agreements, special financing arrangements and clustering small projects into investment portfolios, may be required to focus market interest on energy efficiency. It is important, however, that such interventions address the real preferences and needs of energy users. Actions include:

- fostering voluntary agreements;
- establishing and enforcing building codes and minimum energy performance standards;
- integrating energy efficiency in procurement practices; and
- using government purchasing to stimulate the market for advanced technology.

Ensure Access to Good Technology

The energy-consuming goods and energy supply systems now on the market are generally more efficient than those from the 1970s and 1980s. Still better technologies are going to be needed in the future. This will require concerted research, development and demonstration. In some cases, focused procurement may also be useful. Some of these activities will follow from full cost energy pricing and other market reforms, but additional encouragement and co-ordination by governments may be needed. In any case, the efforts must pay constant attention to the real preferences and needs of users, deployment issues, and communication amongst all pertinent market actors, including manufacturers, users, distributors, energy utilities, business and technical associations, and governments. Actions include:

- encouraging the development, adaptation and diffusion of energy efficient technology;
- improving district heating systems; and
- expanding the use of combined heat and power.

Develop and Maintain a Supportive Institutional Framework

Because energy-use is so dependent on the infrastructure that societies create for themselves, it is vital that energy efficiency principles be embedded into sectoral policies on housing, commercial buildings, industry and transport. Stand-alone energy-focused policies themselves will do little to influence the millions of everyday energy use decisions. Collaboration with the relevant authorities is required to ensure that their policies reflect energy efficiency objectives. However, there remains a need for energy-focused organisations. They link energy efficiency activities across sectors, join energy-specific efforts to environmental policy and assess overall supply/demand concerns. This expertise is needed at local, national and regional levels. There is no unique way to manage energy efficiency policy. In some countries, the energy efficiency body is centralised within a department or ministry. In others it is an independent public agency, or a consortium of public and private entities. In whatever form, an energy efficiency organisation needs well-trained experts who can provide impartial expertise. Actions include

- integrating energy efficiency in sectoral policies; and
- ensuring the availability of impartial expertise.

Act to Ensure Continuity

Large-scale energy efficiency improvements take time and require a policy approach that is transparent, consistent and steadfast. Uncertainty and ambiguity in policies drain energy, effort and resources away from meeting goals. Governments must be clear about their goals and their expectations of individuals and businesses in attaining them. Their policies need to be based on strong analytical bases, including thorough assessments of what policies and measures have been successful in the past and why. Energy demand and efficiency issues need to be fully integrated into energy policy, as well as sectoral policies, because energy security and environmental issues cannot be solved by an energy supply approach alone. The policies should also be routinely monitored, evaluated and revised in order to keep them tuned to changing consumer demands, technologies and other parameters, and to bolster confidence in their effectiveness. Further, harmonisation efforts and international co-operation are needed to join market forces and strengthen policy measures to speed technology dissemination. In addition, governments and institutions need to show leadership by demonstrating the viability and desirability of energy efficiency investments and behaviour in their own operations. All of these factors help to establish an energy efficiency ethic that functions even in times of political change.

Actions include:

- establishing policy clarity;
- demonstrating leadership;
- implementing effective evaluation and monitoring techniques; and
- strengthening international collaboration.

There is no doubt that energy efficiency has a key role to play in leading society toward a more sustainable path. It is equally clear that well-developed markets and well-designed policies are necessary to realise the full potential of energy efficiency opportunities. The experiences reported upon here can assist policymakers in their efforts to develop and implement effective energy efficiency policies.

RENEWABLES: STRATEGIES FOR MARKET ACCELERATION

The Market Potential for Renewable Energy

Strategies and measures for accelerating the market deployment of renewable energy technologies have received greater attention in recent years as renewable energy markets accelerate in developed countries and are beginning to show great promise in developing and transition economies. This is a positive trend in the perspective of sustainable development, as the use of renewable energy can further all three dimensions of sustainability. In addition to diversification and the reduction of supply risk, renewable energies can mitigate environmental risks, such as acid rain or climate change. In the social dimension, renewable energy can contribute to regional and agricultural development. In economic terms, greater use of renewables at home can lead to the development of export markets for energy technologies. In addition, renewables are modular technologies – flexible and comparatively simple.

Renewable markets are growing rapidly. Over the past several years, wind and PV markets have grown at an annual rate of over 20 per cent. Geothermal, biomass and hydro have had somewhat lower growth rates. However, experts believe that many new markets could sustain even higher rates were it not for a number of barriers, including:

- The inertia of entrenched conventional technology industries.
- Limited awareness that renewables contribute to different dimensions of sustainable development, through additional energy security, reduced energy supply and cost risk, and environmental benefits. An important constraint has been the difficulty of instituting policies that internalise those benefits and stimulate renewables markets.
- Continuing and exclusive use of delivered price per kilowatt-hour to compare energy options, which fails to take into account

the secondary-benefits of renewables, and the resulting perception that renewables are comparatively costly.

- The only recent development of marketable products based on renewables and, in consequence, the relative lack of awareness by consumers of those products.

Ultimately the success of renewables will hinge on their acceptance by consumers as "value-competitive". This may occur in a niche market, in more broadly based market for a specific service, or in a general energy market. It is necessary to specify the target market for each renewable and consider the specific measures that will effectively launch it in competition with conventional alternatives.

On the assumption that appropriate policies and measures are enacted, the IEA's World Energy Outlook 2000 suggests that non-hydro renewables will be the fastest growing primary energy source in world energy mix, with an annual growth rate averaging 2.8 per cent over the outlook period. However, due to the small base from which this expansion begins, the share of renewables would reach only 3 per cent of primary energy by 2020 from the current 2 per cent share worldwide. Most renewables are expected to be used in the power generation sector of IEA/OECD countries. Preliminary analysis of the market impacts of planned policy changes supporting renewables in OECD countries indicates more dramatic growth of 8.6 per cent [80].

The costs per unit of output of most renewable technologies are anticipated to decline in the future faster than the costs of fossil fuel technologies. This is due to the relatively greater untapped technological potential of renewables and the effects of experience with new technologies that leads to declines in cost as manufacturing scale grows. "Learning investments" (discussed further below) are made by governments to encourage the development and deployment of new technologies and by the private sector to improve their market position.

80. Growth projections in renewables markets tend to vary, in part because of the use of differing definitions of "renewable energy", established at different times by different agencies. For example, hydro, waste biomass, and geothermal are sometimes not considered renewable.

Box 9

The Role of Governments in Renewables Markets

All IEA governments promote renewable energy in some form or other. Many national and international organisations are also in a position to contribute to these strategies and some are already doing so. Yet efforts so far have been fragmented; lessons from experience have not been adequately synthesised into forms useful for decision-makers; and the development of policy recommendations and decision tools has lagged the needs of decision-makers in both the public and private sectors. Furthermore, among investors confidence in the technical and financial viability of renewable energy technologies is only gradually taking shape.

Technology transfer in the area of renewable energies has recently been given a higher policy priority through an initiative of the G-8 countries at their meeting in Okinawa in July 2000 – they have created a Task Force for the promotion of renewable energies, in which the IEA is involved. The communiqué from the meeting states that "... the increased use of renewable energy sources will improve the quality of life, especially in developing countries..." and advocates "working together and with existing institutions to encourage and facilitate investment in the development and use of sustainable energy". The IEA is contributing to these international efforts through information dissemination the elaboration of "green markets" scenarios and the study of the benefits and impacts of renewable energy market growth.

There is a need to analyse and quantify the benefits that renewables provide, beyond the energy they produce, in terms of energy security, reliability and environmental performance. Although many existing studies quantify these benefits, there is a need for application-specific and sector-specific information that can be used by policymakers and other decision-makers.

To facilitate more rapid deployment of renewables the advantages of using more renewables need to be fully reflected in energy prices. In the increasingly liberalised energy markets of IEA countries competitive pressures force utilities toward the lowest cost option within accounting systems that do not recognise the value of clean energy and the benefits of off-grid renewables on system performance. Risk management and financing issues will become increasingly important, especially in the context of sustainable development. Most of the near-term growth of renewable markets in IEA countries is expected to be achieved in response to these new forces in the utility industry by way of various forms of incentives.

Policies to stimulate renewable markets in a number of IEA countries have been based on portfolio quotas or feed-in tariffs. These have generally led to the increased use of renewables, although most of this growth has been within the borders of those countries. Not all such policies have been entirely efficient and successful. Current levels of investment are insufficient to close the gap between existing growth rates and medium-term environmental goals. To achieve higher rates, additional investments are needed and national policy frameworks need to be strengthened.

The potential market for renewables in developing countries is even larger than in IEA/OECD countries. Contributing factors are (1) climate, (2) untapped potential for geothermal, wind or hydro and (3) geographic size and undeveloped power infrastructure, which makes off-grid solutions more competitive. Investment in renewables might also be supported by measures intended to draw developing countries into the process of implementing the Kyoto Protocol, such as the Clean Development Mechanism under the UNFCCC (see Chapter 8).

It must also be recognised, however, that the challenges are greater in developing countries. They have fewer resources and tools to expand their markets and are caught between competing priorities. Nevertheless, substantial commitments have been made by an increasing number of developing countries as they recognise the benefits of renewables for development, social justice and quality of life. The lack of energy sector infrastructure in many developing countries may give them the opportunity to leapfrog to more advanced energy technologies, including renewables.

Policies and Measures to Accelerate Renewable Deployment

Specific Instruments for Different Markets

Governments can employ a number of strategies to improve the competitiveness of renewables through the clarification of policy frameworks, through "learning investments" (see below) and through internalising the benefits of renewables for energy security and environmental performance.

Incentives such as the "green pricing" of renewable-generated electricity, carbon or other environmental taxes, and giving risk guarantees can add to the financial returns of renewables projects. Monetising and incorporating the benefits of renewables into policy frameworks will signal the private sector that the government recognises that benefits accrue not only to project developers, but also to society.

It is useful to define three primary commercial markets for renewables – utility power, distributed markets and rural off-grid markets – and to consider specific measures that can be applied in each particular market. (Table 9 provides an overview of the different measures proposed for the three markets under study.)

Utility Power Markets

In both OECD Member and non-Member countries, power production from wind, biomass, bagasse, geothermal or hydro resources is already commercially competitive, or is becoming competitive, with conventional sources for bulk power generation.

Other technologies, like central-station solar PV, solar thermal hybrids with gas turbines, and integrated biomass gasification/gas turbine power plants, are not yet commercial, but show promise for the near future.

Grid-connected wind turbines have provided ancillary benefits in terms of employment and technological development. In Denmark it is estimated that about 9000 people were employed directly and indirectly in the wind energy industry in 1995. Denmark had more than 50 per cent of the global market for wind turbines in the late 1990s. In India, the strong growth of the wind industry in the 1990s led to the development of more than 20 joint-venture wind turbine manufacturers. Many of these ventures have developed state-of-the-art technology and have exported turbines to foreign markets[81].

In utility power markets, sustained growth in the use of renewable energy could be greatly facilitated by establishing a transparent framework and rules governing competition and support during the market entry phase, and competition within the mature market. This would have implications for dispatch and pricing. For various renewable power technologies – notably solar, wind and geo-thermal plants – regulatory frameworks that allow fair competition for electricity generation by independent power producers, including power purchase agreements and a transparent and stable tariff-setting regime, are an essential first step. Removing subsidies for conventional generation also needs to be considered in relation to developing renewables markets[82].

Beyond this stage-setting work, substantial experience has been accumulated with specific market development mechanisms to promote the expansion of renewable energy in utility power markets. This experience can be synthesised for IEA/OECD countries and adapted for conditions in non-Member countries. Both failures and successes to learn from exist. Examples include

---

81. *IEA (1998),* The Evolving Renewable Energy Market; *Eric Martinot (1998), "Monitoring and Evaluation of Market Development in World Bank/GEF Climate-Change Projects".*

82. See *IEA (1999),* Looking at Energy Subsidies: Getting the Prices Right.

the Non-Fossil-Fuel Obligation in the United Kingdom, the electricity feed laws in Germany and Spain, and renewable energy portfolio standards that are applied at the state level in the United States[83]. Assessment of the effectiveness of such strategies and comparisons of the extent to which they are used can lead to an understanding of appropriate measures for differing national conditions.

Besides utility-regulation and power-purchase mechanisms, other categories of policies which encourage renewables projects include:

- carbon taxes (notably in Sweden);
- emission taxes (in some U.S. states);
- investment tax credits (adopted in India where they led to a boom in the wind power industry);
- production tax credits;
- wheeling policies for small power producers of renewable-energy-based power (adopted in India);
- green labelling/certificates and green power marketing (in the Netherlands and some U.S. states);
- voluntary agreements by utilities to install renewable energy capacity (Japan);
- mainstreaming renewables into national energy policy;
- technology testing and certification procedures and institutions.

Distributed Markets

In IEA Member countries, applications of renewables in utility-distribution systems are appearing, with generators that range from kilowatts to 30 MW. These can be based on PV, various biomass technologies and small installations of hydro, wind, geothermal and solar, according to the availability of each resource. These

83. Dan Shepherd (1998), "Creating a Market for Renewables: Electricity Policy Options for Developing Countries".

applications may involve generation peak shaving, installations that strengthen the distribution system, building-integrated systems, avoided upgrades to distribution systems, consumer self-generation, and conjunctive uses with other non-baseload or seasonal sources.

Policy experience in distributed markets is only slowly emerging, primarily in regard to the promotion of early market-learning investments. For example, experience with "net metering," where a consumer can sell self-generated power back to a utility at the same cost as purchased electricity, is emerging in Japan and the U.S. Another emerging policy innovation is the reshaping of traditional utility least-cost planning, which historically has focused on generation costs only, to require a broader optimisation of combined generation, transmission, and distribution costs. Such a new paradigm for utility planning would show many commercial opportunities for distributed generation based on renewable energy sources. These and other policies may spur distributed markets for renewables as the utility system is transformed from a uni-directional grid to a two-way network[84].

In several countries, aggressive policies are promoting "roof-top" photovoltaic systems, particularly in the U.S., Japan, and Germany. If dispersed on a large-scale, these applications could displace significant quantities of local fossil-fuel generation particularly for peak periods, thus reducing local air pollution. In addition, employment benefits in the installer and service industries are expected to be large. In areas of developing countries with low service availability, distributed generation can offer households, industry and public facilities a more reliable power supply.

In general, electric power reform will strongly affect the evolution of markets for distributed generation. Distributed markets must be explicitly considered in power sector reform policies if renewables are to compete. Where utilities are regulated according to the traditional manner and rewards are based on the

84. *Seth Dunn and Christopher Flavin (2000), "Sizing up micropower".*

amount of capital equipment owned by the utility or the amount of power produced, regulatory schemes need to be modified so that distributed markets are properly evaluated.

Fair rules also need to be incorporated into utility "unbundling" of generation, transmission and distribution so that small producers using renewable energy sources can compete fairly with the new utility entities. All of these policy measures need to be incorporated in the process of utility reform; changing the rules "after the fact" will elicit strong resistance from the players already operating according to an established set of rules. A key need is to produce the knowledge tools for utilities and their regulators to help ensure that these steps are taken.

Rural Off-grid Markets

Particularly in non-Member countries, applications of small wind, small hydro, biomass, bagasse, and PV in village mini-grids or for stand-alone household systems are proving commercially competitive with conventional alternatives. Solar PV home systems can eliminate or reduce the need for candles, kerosene, LPG, and/or battery charging. Direct economic benefits to rural households include the avoided costs of battery charging and LPG or kerosene purchases. Other significant benefits include increased convenience and safety, improved indoor air quality, better reading light than kerosene lamps for reading, and reduced CO_2 emissions[85].

In off-grid rural markets in developing countries, renewables are more service providers than electricity producers. It is not the electricity *per se* that is valuable, but what it produces, such as light, heat, cooling or water pumping. Thus a critical concern for the development of a successful large-scale rural renewables deployment programme is its linkage to national rural sustainable development strategies. In this context grid extension, energy-

85. A. Cabraal, M. Cosgrove-Davies, and L. Schaeffer (1998), "Accelerating Sustainable Photovoltaic Market Development"; Eric Martinot, Anil Cabraal, and Subodh Mathur (2000), "World bank solar home systems projects: Experiences and lessons learned 1993-2000"; Dan Kammen (1999), "Promoting Appropriate Energy Technologies in the Developing World".

service concessions and model village power schemes are of particular interest[86].

Rural electrification policies can more formally and explicitly recognise and incorporate solar home systems and other viable off-grid renewable energy sources into rural electrification planning. Regulated off-grid energy service concessions are one promising way to provide energy services to rural populations without access to electricity. A review of the World Bank's solar home systems projects suggests that concession tariff-setting, bidding and regulation require substantial time and resources and that regulatory models are still being developed.

Regulatory and social models for tariffs, contracts, ownership and financing are important components of village-power applications. Knowledge of these models must be developed in order to facilitate the promotion of village power schemes and adaptation of models to specific local conditions.

Renewables and National Innovation Systems

Renewable energy markets have always progressed as the result of a double set of forces. On the one hand, they have been supported by policy measures motivated by the contributions of renewables to public goods and the offsetting of environmental externalities associated with conventional energy. On the other hand, improvements in renewable technologies and in their management have enhanced their commercial competitiveness. Such improvements have been made possible through sustained efforts in the research and development of renewable technologies.

Research and development (R&D) has played an important role in the development of renewables. Experience with modern renewable technologies began only about 25 years ago. Since then

86. Eric Martinot, Anil Cabraal, and Subodh Mathur (2000), "World Bank solar home systems projects: Experiences and lessons learned 1993-2000".

Table 9

Overview of Measures to Accelerate Renewable Energy

Generic Measures	Utility Power Markets	Distributed Grids	Rural Off-grid
Enact facilitative electric utility policies and regulation.	Independent power producer frameworks and power-purchase tariffs. Regulatory tools like non-fossil fuel obligation (UK), feed laws (Germany and Spain), standards, system benefits funds and renewables portfolio standards (U.S.). "Green pricing".	Consumer net metering.	Rural electrification policies and planning that integrate renewables in serving areas too costly for grid extension. Regulated off-grid energy-service concessions. Model village-power schemes (tariffs, contracts, ownership, financing).
Develop tools and outreach to increase demand and confidence.	Renewable resource assessments. Technology cost and performance information.	Utility distribution planning and analysis tools. Technology cost and performance information.	Market surveys of off-grid households to determine affordability, energy-service needs, and current energy consumption.
Strengthen national systems of innovation.	Technical assistance to wind turbine producers in developing countries to reduce production costs, increase quality. Technical assistance to wind farm developers on siting and operating and maintenance.	Expanded institutional linkages between distribution utilities in developing countries and PV R&D personnel and utilities in OECD countries.	Development of technical standards and certification institutions for solar home systems in developing countries.
Increase financing sources and opportunities.	Develop financial and institutional intermediaries (like IREDA in India) that can finance renewable energy projects. Outreach to commercial financiers to familiarise them with technologies and increase confidence in investments.	Consumer financing mechanisms. Support development of appropriate business models.	Development of microfinance and/or fee-for-service financing models in rural areas. Facilitate private sector funds development.

the considerable investment made in R&D has demonstrated the potential and proven the technical viability of renewable technologies; in some cases, it has led to successful and rapid market entry strategies. However, to reach their full potential, renewables require additional investment to improve performance. To make a more significant contribution to the energy portfolio, a new generation of advanced technologies must be developed.

Information on energy R&D spending by IEA governments indicates that the share of resources devoted to research on renewable energy technologies grew from 6.1 per cent in 1990 to 8.2 per cent in 1998. Favoured options in the allocation of funds are solar electric (PV), biomass and wind. In real terms, however, this has meant only a slight increase in overall funding, which grew from US$ 549 to 586 million (at 1998 prices and exchange rates) from 1990 to 1998. However, considering that total public budgets for energy R&D decreased in that period, the increased level of funding for renewable technologies is an indication of the political will to expand the markets for renewables.

Consistent with that, more attention is now being paid to "national systems of innovation", a term used to refer to networks of institutions that initiate, modify, import and diffuse new technologies. J.S. Metcalfe defines such national systems as "the set of institutions to create, store and transfer the knowledge, skills and artefacts which define technological opportunities". National systems of innovation reflect a complex mixture of institutions, public policy, business and social relationships[87]. IEA Member countries have committed themselves to extending the traditional focus of technology development to encompass the encouragement of learning investments and the facilitation of commercial market deployment[88].

87. *OECD (1999),* Managing National Innovation Systems; *J.S. Metcalfe (1995), "Technology systems and technology policy in an evolutionary framework".*

88. *The IEA has at various times published reports that focus on policies and measures for accelerating market deployment. See IEA (1994), Development and Deployment of Technologies to Respond to Global Climate Change Concerns and Scoping Study: Energy and Environmental Technologies to Respond to Global Climate Change Concerns; IEA (1997),* Energy Technology Availability to Mitigate Future Greenhouse Gas Emissions *and* Renewable Energy Policy in IEA Countries. Volume I: Overview; *IEA (1998), The Evolving Renewable Energy Market.*

National systems of innovation can influence a multitude of technologies simultaneously. But each technology may have it's own "network(s) of agents interacting in each specific technology area under a particular institutional infrastructure for the purpose of generating, diffusing, and utilising technology"[89]. These networks may span across national boundaries, but are still subject to the influence of national culture, institutions, and policies. Empirical work suggests that both the overall system and the quality of interconnections within it affect successful innovation[90].

Policy interventions can address particular weaknesses in networks. Creating the knowledge, institutional and administrative infrastructure – often referred to as "capacity building" – is one aspect of strengthening national systems of innovation. More important is the creation of linkages and long-term partnerships across the whole spectrum of technological innovation, from research to commercialisation. This can include considerations of finance, marketing, organisation, training, relationships with customers and suppliers, distribution networks, service, and competitive positioning. Finally, regional and global systems of innovation need to be integrated with national-level systems.

In strengthening systems of innovations three guiding principles need to be observed:

Analyse the system of innovation seen as a whole. Where are its strengths? How does the institutional framework around the processes of innovation and learning compare with that of other countries or regions? Where are the important connections and missing linkages?

- Focus on capabilities to absorb knowledge developed elsewhere, in particular foreign technology and information technology. Introduce new elements and institutions in a way that takes into account the characteristics of the existing system of innovation.

89. *Bo Carlsson and R. Stankiewicz (1991), "On the Nature, Function and Composition of Multinational Enterprise".*

90. *Mark Dodgson and John Bessant (1996),* Effective Innovation Policy: A New Approach, *p. 54.*

- Focus on end-users when shaping market deployment strategies. Give users the competence and resources to develop, select, adapt, and understand new technologies. Understand models of appropriate distribution and service chains and determine how to replicate these models[91].

Targeted Learning Investments in Order to Further Accelerate Markets

"Learning investments", i.e., expenditures that need to be made to bring a new technology to the point of commercial acceptance, are special element in national systems of innovation. They are particularly relevant to renewable energies because the technologies involved tend to be new. Learning investments lead to reduced costs through the accumulation of hands-on experience ("learning by doing"). As individuals and industries gain experience with technologies through a competitive market, they learn to improve technology and reduce implementation costs. For example, experience with wind turbine deployment over the past two decades shows that cost-per-installed-kilowatt has decreased dramatically as manufacturers have improved their designs. In addition, costs per produced kilowatt-hour have decreased even faster, as developers have learned how to better identify good wind farm sites, optimise turbine micro-siting, perform maintenance, and manage power production[92].

Strong evidence across industries shows that experience with supplying technologies reduces prices and that relatively simple, quantitative relationships exists between accumulated experience and price. A familiar example involves computer chips, where cost as a function of cumulative manufacturing volume can and must be predicted quite accurately, as the time from generation to generation is a year or two, rather than decades. Renewable energy technologies appear to share these characteristics.

91. *Bengt-Ake Lundvall, "Technology policy in the learning economy", cited in IPCC (2000),* Special Report on Methodological and Technological Issues in Technology Transfer.

92. *Paul Gipe (1995),* Wind Energy Comes of Age; *IEA (2000),* Experience Curves for Energy Technology Policy *(IEA, 2000); Williams, R.H. and G. Terzian (1993), "A benefit/cost analysis of accelerated development of photovoltaic technology".*

Box 10

Learning Investments: a Graphical Demonstration

Where new technologies cost considerably more than current alternatives, the main barriers to adoption are often those common to infant industries, where with time and scale of production costs can be expected to fall significantly in the future. In such cases, government support can help speed up development and deployment. This is true of many renewable energy sources. They are, for the moment, expensive compared with conventional ways of producing electricity – the price for electricity generated from solar photovoltaic cells is currently about 10 times larger than from fossil fuels[93].

Figure 18 illustrates the problem confronting policymakers. The "experience curve" shows the expected evolution of the unit cost of a new,

Figure 18

Hypothetical Experience Curve for a New Technology without CO_2 Emissions

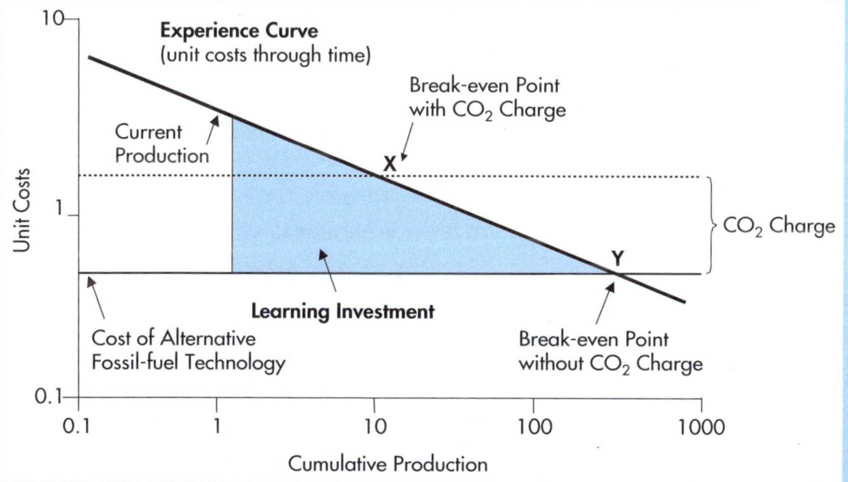

Source: Adapted from Figure 1.4 in IEA (2000h).

93. *IEA (2000h),* Experience Curves for Energy Technology Policy.

no-emission technology as experience with its manufacture and operation accumulates, and as economies of scale begin to be exploited. The two horizontal lines represent the costs of fossil fuel-based alternatives. The upper one incorporates the cost of internalising externalities associated with fossil fuel-based technology; the lower one does not. The break-even points for the new technology, X and Y, represent the thresholds below which the unit costs of the new technology must fall in order for it to compete commercially in the power-generating market.

In order to bring costs down, experience must be gained and market size must expand. Most private investors will be unwilling to take a risk on a technology that will yield no short- or medium-term profits. Policy intervention may therefore be warranted, especially if the risk-adjusted costs of bringing the technology to maturity are small relative to the long-run benefits that will accrue to society from the new technology.

Expenditures on learning investments can be shared between governments and private firms. Public R&D can "seed" the learning process and keep it on track and technology policy focused on market deployment can support its share of learning investments. At some point, private firms will become willing to make learning investments on their own, in combination with public investments. This point, referred to as the "docking point", marks a high degree of technological maturity and industry willingness to invest, but not full commercial viability. For example, analysis by Wene showed that for wind turbines in Germany about two-thirds of the total learning investment of DM 800 million during the 100/250 MW feed law programme from 1991-1997 came from the private sector, while the remainder represented public subsidies[94]. Finally, near the point of full commercial viability, learning investments are no longer needed.

94. *Clas-Otto Wene, Alfred Voss and Tanja Fried, Eds. (2000),* Experience Curves for Policy Making – The Case of Energy Technologies.

The tendency for progressively greater investment by the private sector as costs come down and learning takes place means that public subsidies can be designed to decrease over time and "track" cost reductions. Too rapid a decline, however, can stifle the emerging private learning investments, while too slow a decline can waste public resources. (The German feed laws have been cited as examples of subsidy programmes that do not track the decline in costs).

The challenge for policymakers is to mobilise sufficient private capital during the deployment stage while avoiding the temptation to "pick winners". Deployment policies should be designed to allow competition among technologies that can meet the same objective. Learning investments must be made in multiple competing technologies until the accumulated experience demonstrates which technologies will be commercial winners. An illustration of this is how absorber solar-heating for swimming pools proved superior to collector technology, although learning investments were needed in both before this superiority became apparent.

Conclusion and Proposals for Action

Global strategies for accelerating the market penetration of renewable energies need to engage both IEA/OECD countries and non-Member countries. A key element of the proposed market acceleration strategy is facilitating the linkages between countries with significant renewable resources and a commitment to develop a market for the technology to exploit that resource. These "lead markets" are indicated by the government to encourage the renewable industry, particularly through the establishment of financial incentives. Such lead markets exist in both industrialised and developing countries. Special attention is warranted for developing countries, where the potential for renewable energies might well be greatest.

In order to make renewables widely competitive, it is necessary to develop a policy momentum in a group of countries that have good

Box 11

Information-Provision and Institution-Building to Overcome Market Barriers

Below are several examples of initiatives to develop information tools and partnerships that can contribute to overcoming market barriers and accelerate the deployment of renewable energy technologies.

The IEA Photovoltaic Power Systems Program (PVPS) is conducting several tasks aimed at the removal of technical and non-technical barriers in OECD countries through a variety of deliverables, such as recommended practices, policy surveys, and financing reviews. A recent activity is the development of a recommended practice guide that encompasses policy and planning, financing mechanisms, and institutional development.

The World Bank ESMAP programme has produced a wealth of experience and documentation on applications of renewable energy in developing countries. In particular, several important lessons have been learned about the application of renewable energy technologies in rural areas, such as the need to be "technology neutral" in coming up with the best solutions to meet user needs, developing new delivery and business models, and using subsidies to make rather than destroy markets[95]*. The recently formed a PV Consultative Group will develop public-private partnerships to facilitate the application of PV in developing countries.*

The Technology Co-operation Agreement Pilot Project (TCAPP) in the United States works with five developing countries to prepare frameworks for technology co-operation that consist of priority technologies and actions. Once these frameworks are in place, country teams will design and implement actions to address market barriers, secure donor support for barrier-removal actions, conduct market

95. Andrew Barnett (1999), "A Review of the Renewable Energy Activities of the UNDP/World Bank Energy Sector Management Assistance Program, 1993-1998".

assessments, identify immediate business opportunities for priority technologies and attract private investment.

The Global Environment Facility continues to evolve and document the approaches it takes to remove market barriers to renewable energy technologies and to reduce their long-term costs. Thematic reviews of the experience of the GEF renewable energy project portfolio suggest the types of policies, capacity building, financing mechanisms, and business models that could lead to rapid deployment of renewable energy in developing and transition economies[96].

renewable resources and strong commitments to accelerate learning and market scale. Such a strategy will concentrate and accelerate the learning experience, leading to greater efficiency and reducing overall costs. The strategy should be focused on five key elements:

- Integration of the market, both at the national level to integrate renewables into the country's energy portfolio, and at the international level, focusing on strengthening the technical, financial and business interaction between leading countries.
- Co-operation and co-ordination with multilateral organisations to provide resources and pool competencies for a coherent programme at the international level and to promote national policy frameworks with incentive strategies and investment support.
- Engagement and support of the private business sector to enlarge the industrial support for renewables, and increase the level of private sector investment in technology development and infrastructure development.
- Building confidence in the financial community about the current and future viability of renewable technologies and the policy frameworks being established to sustain commercial markets.

---

96. Global Environment Facility (1996); Eric Martinot and Omar McDoom (1999).

- Understanding the opportunities and characteristics of financing renewables projects and highlighting the gains from investing in renewables (e.g., reduced portfolio risk) in order to increase the public perception of their value.
- Acquiring a better understanding of how necessary learning investments should be made. Co-ordinating learning investments across national boundaries by different public and private organisations so that "learning" is shared and not needlessly repeated.
- Push for public policies and programmes that may require initial subsidies but involve the aim to eliminate the need for subsidies at a later stage.

The main goals of these activities are to increase the scale of deployment and to create confidence in the technologies and markets. Both goals will have the effect of attracting private investments and bringing renewable technologies into the mainstream. Renewables are a potential solution to many sustainability challenges, such as energy security, social and economic development, and perhaps most significantly, environmental protection. Until the full commercial competitiveness of renewables is established, their positive contributions to sustainability justify policies and incentives supporting their use.

POLICIES FOR SUSTAINABLE TRANSPORT

Transport – a Stumbling Block on the Way to Sustainability

IEA countries consume the vast majority of their oil imports and emit roughly one-third of their energy-related greenhouse gas emissions in the transport sector. Transport is thus at the intersection of several different sustainability concerns and is a priority area for policymakers. The three dimensions of sustainability in transport can be described as:

- *Economic dimension*: The provision of adequate, affordable transport options to satisfy society's needs for access and mobility and to move goods. This includes the maintenance of a sustainable level of oil imports (see Chapter 3);
- *Social dimension*: The provision of adequate transport services for all members of society provision in a manner that does not damage the "social fabric" including safety, health, congestion and equal access to services for different groups of the population;
- *Environmental dimension*: The provision of transport services in a manner that does not degrade the environment or hinder people's ability to obtain other needed resources or carry out other needed functions with those resources. A key aspect is the reduction of greenhouse gas emissions.

While it may not actually be possible to meet each criterion completely, the three together provide a framework for defining what transport sustainability looks like and in which direction transport policy needs to move.

By almost any measure, trends in transport energy use and greenhouse gas emissions in IEA countries are currently on an unsustainable path. Fuel use continues to grow steadily, and recent CO_2 increases are among the highest of any sector (Figure 19). Most forecasts indicate that these trends are not likely to change significantly in the coming years without substantial new policy initiatives. Some other indicators of transport sustainability, such as emissions of various air pollutants, appear to be on a better trajectory, with policies in place to yield significant reductions in many countries over the next ten years, continuing the downward trends established in recent years.

Figure 19

Growth in CO_2 Emissions in IEA Countries by Sector, 1990-1997
(Million Tonnes of CO_2)

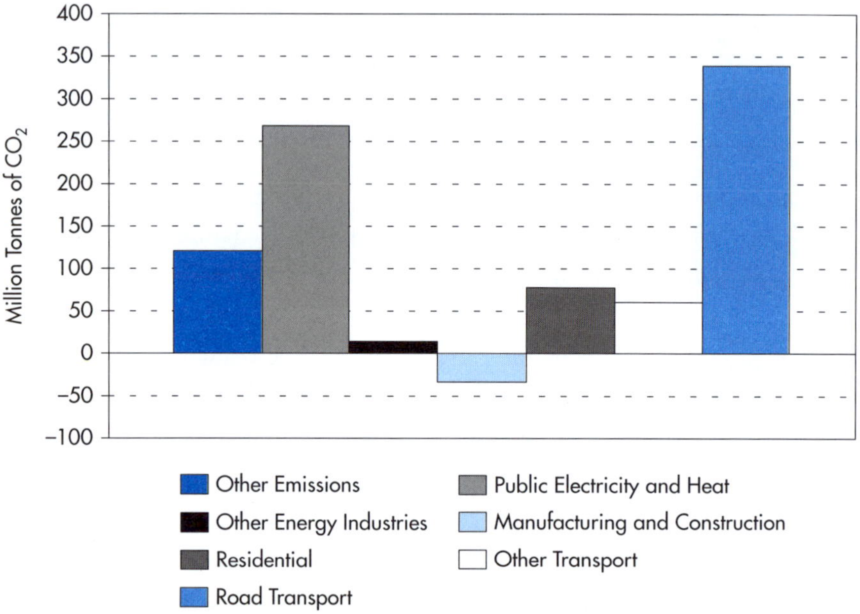

On the other hand, trends in transport activity are rising steadily and show few signs of "saturation" in terms of passenger-km *per capita* that might moderate growth in the future. Transport activity (passenger-km of personal travel or tonne-km of freight) itself is an important indicator of sustainability in terms of its role as the main "driver" increasing fuel use and emissions, as well as its impacts on the environment. This chapter focuses on transport energy use and CO_2 emissions, but it also considers transport activity itself as an important sustainability indicator: energy use in transport is closely tied to how much and how we move people and goods.

Over the long term, a key element of transport sustainability is managing the risks and threats of major disruptions and discontinuities to the transport system itself and to the broader set of human and natural systems that transport interacts with and affects, i.e., nearly everything. In this regard, one of the biggest long-term concerns for transport is its near total dependence on petroleum fuels. Conversely, the petroleum dependence problem in IEA countries has become primarily a transport problem with transport now accounting for nearly two-thirds of their petroleum consumption. As petroleum supplies dwindle and become increasingly concentrated in the Middle East, the risk of supply disruptions and price shocks is likely to increase. Management of these risks, for example through diversification of energy sources, represents a critical aspect of long-term transport energy sustainability (see also Chapter 3).

A second major risk to long-term transport and global sustainability is climate change. Transport is currently running on a "collision course" with a world increasingly convinced of the need to take major steps to reduce CO_2 emissions. It may be possible to meet Kyoto commitments without outright reductions in transport CO_2 emissions over the next ten years. But a major risk for transport is that the longer countries wait before beginning a shift toward significant decarbonisation of this sector, the more likely it is that such a shift will have to happen very quickly and at high cost. The sooner transport is shifted onto a path of reducing CO_2 emissions, the less likely the chances of a worst-case scenario (see also Chapter 8).

These long-term risks suggest that several of the sustainability issues outlined in previous chapters are important in developing a sustainable path for the transport sector. They include:

- improving energy efficiency (see Chapter 5),
- increasing the potential for renewable energies (see Chapter 6),
- expanding access to energy technologies as well as more rapid development and deployment of advanced technologies (see Chapter 9),
- developing strong yet politically acceptable policies to shift the direction of transport activity toward a more sustainable path (see below).

Indeed it may be necessary to actually lower, or at least reduce the rate of growth of, the "energy service" itself — vehicle kilometres of personal travel and tonne-kilometres of freight — in order to achieve a sustainable path.

Not all choices for future transport systems will affect all aspects of sustainability equally or even in the same direction. While the dominant or direct environmental impact of energy use in transport is CO_2 emissions, the types and amount of energy used also affect emissions of other pollutants, such as NO_x and SO_2. Decisions that are taken to reduce CO_2 emissions could in some instances increase the emissions of other pollutants. Similarly, while energy use does not directly affect transport's impacts on the natural environment such as habitat destruction the allocation of ever-greater levels of energy to an ever-expanding transport system may itself be unsustainable.

An important element of global transport sustainability is the manner in which transport is handled in developing countries. Non-Member countries are expected to increase their levels of transport and their transport energy use at a much greater rate than IEA countries over the coming decade and beyond. In the *World Energy Outlook 2000* Reference Case, non-OECD transport energy use is projected to grow by 128 per cent between 1997 and 2020, while OECD growth is projected to be about 41 per cent.

Transport's central economic role and its strong influence on daily life have made rapid changes in this sector difficult to achieve. Its relatively weak responsiveness to energy price movements and the slow turnover of its infrastructure make it both crucial and difficult for policymaking to improve sustainability (see also Chapter 2). Relying on recent IEA efforts to elucidate future trends and potential alternative scenarios for transport, this chapter provides a quantitative look at where transport is headed and what steps appear necessary to change this direction and put transport on a sustainable path.

The Future of Transport: Present Trends and Policies

Despite recent policy efforts such as the EU voluntary fuel economy agreement, the Reference Scenario for transport of the IEA *World Energy Outlook 2000* expects that over the next two decades energy demand in transportation will climb faster (at 2.4 per cent per year) than in any other end-use sector. By 2020, transport is likely to account for more than half of global oil demand and roughly one-fourth of global energy-related CO_2 emissions. These shares increase steadily over the outlook period in the OECD area (Figure 20) as well as globally. Oil consumption in transportation has thus become a major policy concern in the context of both increasing oil-import dependence and rising CO_2 emissions.

Among the main transport-energy trends, both historical and as expected in the Reference Scenario, which includes the effect of current policies (see discussion below), several central findings emerge:

- Projected energy demand growth from transportation slows considerably in Europe and substantially in the OECD Pacific area. In North America, the past trends continue.
- Activity growth is slowing somewhat — most in OECD Pacific, less in OECD Europe, least in North America, and less for freight than for passenger travel. Europe's projected road-freight is still growing significantly.

Figure 20

Percentage Share of the Transport Sector in OECD Oil Demand

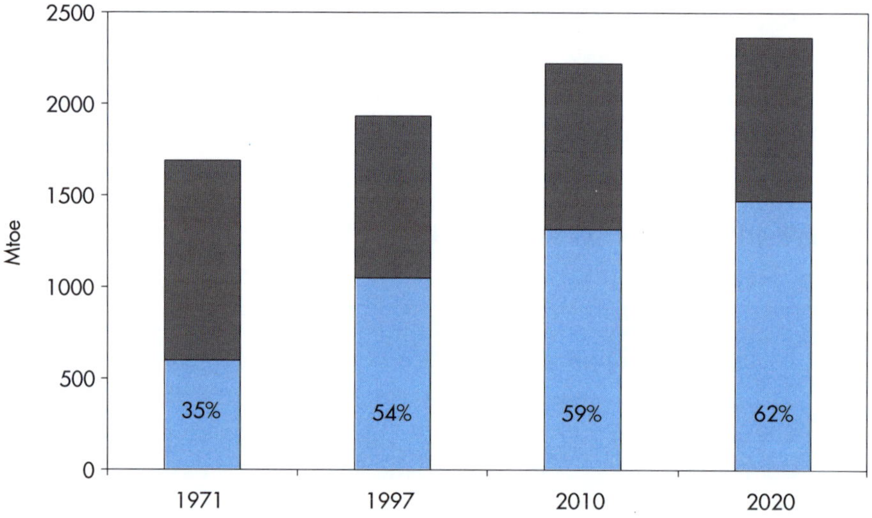

- Without policy intervention, gains from fuel-intensity improvements in passenger transport do not continue. In OECD Pacific and OECD Europe, policies foster continuing improvements, although they offset only a quarter of the fuel demand increase from activity growth.
- In freight, modal shifts will continue to raise average fuel demand per ton-kilometre. Road freight and, increasingly, aviation freight gain shares in most regions. Small commercial trucks and vans contribute about half the energy demand growth from road freight.
- Aviation accounts for almost a quarter of the projected increase in fuel demand to 2020. Fuel-intensity improvements in this mode offset only a quarter of its growth from expanding activity.
- CO_2 emissions from transportation will continue to grow significantly — by more than 60 per cent until 2020 compared with 1990 in each region, under the reference-scenario policy

assumptions. Even as early as 2010, the figures reach 44 per cent in North America and Europe and 48 per cent in OECD Pacific.

The overall picture for the transport sector as provided by the Reference Scenario of the IEA's *World Energy Outlook 2000* is one of continued strong growth in both oil consumption and CO_2 emissions from transport as shown in Figure 21 below.

Figure 21

World Energy Outlook 2000 *Forecast of OECD Transport Energy Use and CO_2 Emissions*

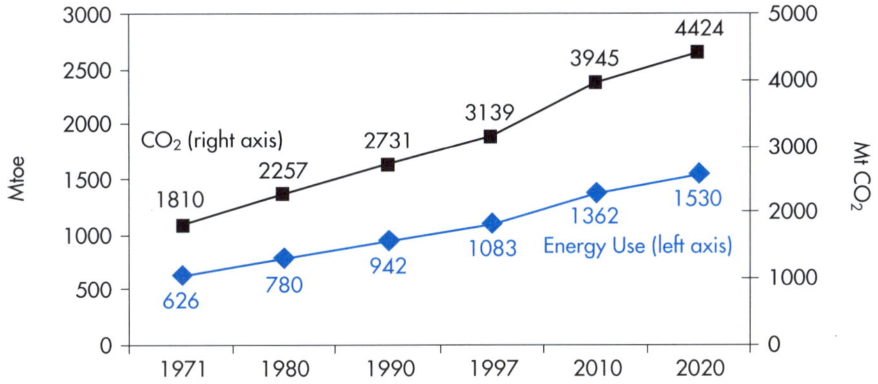

Current Transport Policy Efforts in IEA/OECD Member Countries

The Reference Scenario of the IEA's *World Energy Outlook* contains a number of transport policies that are currently implemented or announced in IEA/OECD Member countries. The projected trends thus constitute much more serious risks to sustainability as they already include several policy measures, highlighting the magnitude of the challenge of rendering the transport sector sustainable.

The recent IEA publication *The Road From Kyoto: Current CO_2 and Transport Policies in the IEA* reviews recent policymaking efforts in Denmark, Germany, the Netherlands, Sweden, the United Kingdom, the United States and the European Union. The report finds that each country has identified potential for reducing or restraining transport-related emissions, but none can expect to achieve reductions sufficient to offset the growth of transport activity in the near term.

The review finds that policies affecting carbon emissions from transport tend to be complicated and hard to implement quickly. No country has yet managed to put in place more than a small number of the measures that were under consideration in the mid-1990s. Some of this lag can be attributed to slow political processes and to political ambivalence. But even when approved, such policies can take several years to implement.

Most of the countries studied have linked CO_2 policies directly to a comprehensive long-term transport policy reform, with an emphasis on "getting the prices right". Other measures have focused on reducing the growth rate or the level of personal vehicle travel and on improving traffic conditions. These policies typically are driven more by congestion and nuisance concerns than by CO_2 reduction concerns. Transport policy reform can also include widespread reforms in the way transport services are priced and taxed, the internalisation of environmental costs and integration of infrastructure across modes.

Perhaps the most important technology-oriented policy measure now in place is the voluntary agreement for emissions reductions in new cars, between the European Union and ACEA, the European Association of Car Manufacturers. This EU-ACEA agreement, if successful, could reduce emissions from cars by as much as 15 to 20 per cent below trend by 2010 for new cars, with even greater reductions after 2010[97].

The largest CO_2-related research project in the public sector is probably the U.S. Partnership for a New Generation of Vehicles

97. *Under the assumptions of the* World Energy Outlook *this translated into a reduction of fleet emissions by 7-9 per cent by 2010 and by 13-17 per cent in 2020 when the more efficient new cars will have fully penetrated the vehicle stock.*

(PNGV). PNGV may well represent the most ambitious – but also among the most uncertain — of the elements in the current U.S. basket of transport, energy and CO_2 policies. Detroit can technically build cars with much higher fuel efficiencies and a number of prototypes were already demonstrated by mid-2000. However, whether auto makers can produce vehicles in the short run that are highly fuel efficient, while maintaining or improving safety, performance, emissions and price remains to be seen.

Measures to encourage modal shifts and non-motorised transport modes as well as measures to encourage less personal travel form part of several European programmes. Improved transport services are expected from the privatisation of railways, but the ultimate effect is unclear. Land-use planning will complement transport policy reforms, but it is difficult to predict how much this will reduce travel or induce modal shifting. In the European countries studied, infrastructure expansion and investment are increasingly tilting toward non-road or inter-modal facilities. Such efforts should at least slow the erosion of the share of low-CO_2 modes. All measures taken together might lead to 5-10 per cent lower emissions than otherwise by 2010, mainly by reducing the distance driven in cars and trucks – a modest slowdown that is already reflected in the growth predictions of the IEA Reference Scenario.

More Ambitious Policies Toward Sustainability in the Transport Sector

As mentioned, recent policy initiatives in IEA/OECD countries, as outlined in *The Road from Kyoto* are already reflected in the *World Energy Outlook* Reference Scenario presented above. These initiatives clearly appear to be insufficient to significantly "bend the curve" of fuel use and CO_2 emissions over the next 20 years. In order to make tangible progress toward sustainability in the transport sector, significant additional measures will be required. While the recent past does not suggest that it will be easy to implement even stronger policies in the future, some possible avenues exist.

For the *World Energy Outlook 2000*, the IEA developed an alternative scenario that identifies several additional measures that appear to be among the most politically feasible for many countries over the next few years. For a "pool" of such potential future policies, we draw on the recent draft report from the *Policies and Measures* project that includes an analysis of a wide range of potential policies in transport[98].

Fuel Economy

The IEA *Policies and Measures* study estimates that there is considerable potential for cost-effective improvements in light-duty vehicle fuel economy in the next ten years and beyond[99]. Using only technologies that are already present on some car models it is estimated that reductions in new car fuel intensity on the order of 25 per cent are possible by 2010. After 2010, introduction of some advanced technology vehicles with radically improved engine and drive-train efficiency (advanced hybrids and fuel cell vehicles) could provide significant additional fuel economy improvements.

The existing EU voluntary agreement and the Japanese "Top-runner" programme are policies already in place that appear likely to help achieve much of the available cost-effective potential by 2010[100]. In North America, no current policy appears to be encouraging the market to take advantage of these technologies. Without additional policy efforts, much of the available technology may be used to increase vehicle size, weight and horsepower, while holding fuel economy constant, as has occurred over the past ten years.

Several types of policies could be implemented to help maximise the fuel economy benefit of existing technology. In addition to voluntary agreements and "top-runner" style policies, fuel economy-based fees and rebates (also called "feebates") and

---

98. *IEA (2001 forthcoming)* Policies and Measures: Transport Volume.

99. *IEA (2000)* Fuel Economy Improvement: Policies and Measures to Save Oil and Reduce CO_2 Emissions *issued at the COP-6 meetings at The Hague, November 2000.*

100. *This Japanese policy, among other things, identifies the most fuel-economic vehicles in each market class and requires all new vehicles to approach that level over time.*

market incentives for the adoption of advanced technologies, could play useful roles in North America and elsewhere.

Alternative Fuels

In the short and medium term, few OECD countries consider alternative fuels to be an important option for mitigating energy-security concerns and greenhouse gas emissions. Many alternative fuels do not offer significant lifecycle greenhouse gas-reduction benefits, especially when derived from fossil resources. In addition, most alternatives would be expensive and require dramatic levels of risky investment.

In the longer term, fuels derived from cellulosic feedstock (ethanol or methanol produced in advanced biological conversion processes) could bring lifecycle greenhouse gas emission reductions beyond 80 per cent compared with fossil fuels. They appear to have a strong cost-reduction potential, using advanced feedstock production and conversion technologies that might become available in ten years. However, it is questionable — even in the longer term — whether they are the best use of scarce biomass resources.

Another potentially important alternative fuel option is the fuel cell vehicle running directly on hydrogen fuel. With renewably derived hydrogen, this technology could provide near zero life-cycle CO_2 emissions. In the 2000-2010 time-frame, however, hydrogen fuel cells are not likely to be produced in large numbers. And those fuel cells that are likely to be fuelled by gasoline or methanol with on-board reformers to extract the hydrogen from those fuels are unlikely to offer any significant CO_2 reductions compared to advanced conventional technology.

Policies to encourage the future development of the needed vehicles and fuel supply systems are complex and could be expensive. The forthcoming *Policies and Measure* report on transport includes carbon-based fuel taxes or outright subsidies for low-carbon fuels. Other measures are direct investment in fuel supply infrastructure and co-ordination of vehicle production and sales with compatible fuel sales in concentrated areas to help the early development of market niches.

Reductions in Growth of Motor Vehicle Travel

Measures that aim to restrain motor vehicle transport volume and to shift it to more environmentally benign modes have begun to play an increasingly important role in long-term policy planning. Sustainability concerns other than greenhouse gas emissions, such as congestion, urban sprawl and quality of life, are often the key motivations to implement such measures. One problem with demand restraint policies for national governments is that many of them need to be implemented on regional or local levels[101]. In order to implement successfully programmes that yield significant reductions in personal vehicle travel, broad packages of policies will be necessary to discourage vehicle travel and provide alternatives.

Hitting Everything: Fuel Taxation

Fuel taxation plays an important role in all the policy issues surrounding the transport sector. For instance, higher prices at the pump provide incentives to choose more efficient cars. Increases in taxation can help limit rebound effects. Differentiated fuel taxes are also widely used to support or limit the use of specific fuels. Finally, fuel-tax increases restrain transport activity and have an implication for the relative competitive advantages of different modes,

Yet one should not overstate the role of taxation, especially as an isolated measure. The response to fuel-tax increases is very limited in the short term, although somewhat greater in the long term. Moreover, considerable uncertainty surrounds long-term price elasticities as a measure of price responsiveness. The values vary over time and depend on the estimation approach, the mode and the region. In many countries fuel taxes are already at levels

101. Demand-oriented policies can encompass pricing of different transport activities, transport supply, investments in infrastructure and rolling stock, regulation and restriction of traffic, urban and land-use planning.

that are not politically popular and significant additional increases in the near future would be problematic. In the future, it is likely that more selective pricing mechanisms, such as km-charges and toll-rings, will find use.

A First Step Toward Sustainability: the *World Energy Outlook* Alternative Scenario

Based on the above policy options, the alternative transport scenario developed for the *World Energy Outlook 2000* crafts a package of policies that are plausible and realistic but not yet enacted or announced. The scenario thus reflects likely, near-term policymaking over the coming years in line with what is currently under discussion. The package outlined is not a radical departure from today's policymaking, rather its prolongation and intensification, but it still could make a significant change in projected energy use and CO_2 emissions from transport. To allow for some lead-time, the alternative scenario assumes that the implementation of such additional policies starts in 2005 [102].

The Reference Scenario assumed that automobile fuel intensity (energy consumption/vehicle km) would fall in the coming decade by 24 per cent in Europe and 17 per cent in Japan. The alternative scenario assumes that beyond 2010 an extension of these trends to 2020 will yield an overall 46 per cent improvement in Europe and 30 per cent in Japan relative to 2000 levels. In addition, fuel economy targets are put in place in North America that achieve a 14 per cent improvement by 2010 and a 31 per cent improvement by 2020. The primary effects by 2010 would be in North America, since this is the only region without any targets in the Reference Scenario. After 2010, significant improvements in new light-duty vehicle fuel economy are also assumed to occur in all three regions.

102. In order to make the analysis manageable, only Western Europe, Japan and OECD North America were studied in detail. The policies taken into account are a fuel tax increases equivalent to US$ 95 per tonne of carbon and fuel efficiency increases in all regions. In addition are assumed the introduction of "low-carbon" fuels in North America, urban car restraint, high-speed rail expansion and truck km-charges in Europe and urban road pricing, improved logistics and the expansion of public transport in Japan.

In terms of alternative fuels, the expansion of the cellulosic ethanol market in North America is assumed with the 'low-carbon' fuel starting production in 2010 and expanding quickly to attain about 11 Mtoe (5.9 billion gallons per year) in 2015 and 27 Mtoe (14.2 billion gallons) in 2020. Under the alternative scenario, 6 per cent of gasoline use and about 3 per cent of total transport energy use under the additional policy case would be replaced by ethanol in 2020. Depending on the initial cost (assumed to be twice that of gasoline at US$ 2 per gallon) and the cost reduction due to the expansion of production, the cost of ethanol in 2020 could already be lower than the comparable gasoline price[103]. The effect on CO_2 emissions would still be modest, at about 90 million tonnes of CO_2, or 4 per cent of the CO_2 emissions from transport.

Concerning changes in transport activity and the modal mix, the *World Energy Outlook* alternative scenario is restricted to assessing what impact a specified change in travel could have on energy demand and CO_2 emissions. These changes are likely to be driven mainly by policy initiatives aimed at objectives other than CO_2 emission reductions[104].

Finally, along with the sector-specific measures, a tax of US$ 95 per tonne of carbon is added across all regions and fuels and phased in progressively between 2001 and 2010 (comparable to the results of the emissions trading simulation described in Chapter 8). The total energy-demand reduction does not vary greatly across the regions. Higher relative fuel-price changes in North America are balanced somewhat by lower assumed elasticities compared with the other regions. The overall reduction in energy demand calculated with this elasticity approach is around 4 per cent across the OECD area.

Important synergies and some overlap occur between the different measures examined in this policy case. This is particularly true for fuel taxation. Its effect on fuel intensity could overlap with fuel-

103. *If the learning rate were higher than 7 per cent, ethanol would be become competitive with gasoline well before 2020 even without the increases in crude oil prices underlying the Reference Scenario.*

104. *For a more detailed discussion of these measures, refer to the alternative fuel case exposition in the* World Energy Outlook 2000.

consumption regulation, and transport-demand reductions or shifts assisted by fuel price changes could form part of demand-side policy packages. As a consequence of such overlaps, the combination of policy measures has been assumed to have an impact *less than* the sum of the results of the individual measures.

Figure 22 presents in one graph the results of the alternative scenario, the Reference Scenario and the historical data for 1971-97. While in the Reference Scenario energy demand growth over the next 20 years is forecast to be about equal to the historical energy increase between 1970 and 1997, it is about one-third lower in the alternative scenario. Although significant, the lower growth is not enough to stabilise energy consumption or CO_2 emissions by 2020.

Figure 22

OECD Reference and Alternative Case Forecasts

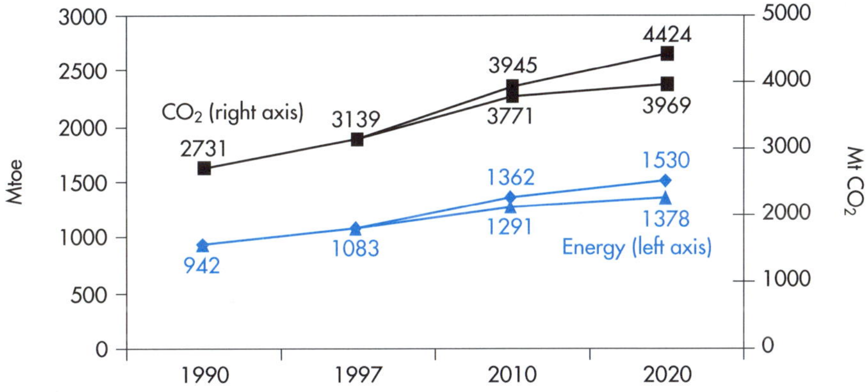

In the alternative scenario, most of the reductions in the growth of energy demand and CO_2 emissions come from cars. Their energy-demand growth in 1997-2020, would be cut by more than half compared to the Reference Scenario. This is due largely to fuel economy improvements in North America, since this region experiences negligible improvement in the base case. Using alternative fuels in North America would reduce CO_2 by about 90 Mt CO_2 in 2020 (but by none in 2010). Some growth reductions are also achieved in trucking and aviation, in reaction to the demand-side policies and the carbon tax. In all regions, fuel-price increases and demand changes in the policy case have some effect, but do not dramatically alter travel behaviour.

The results of the alternative scenario allow the following conclusions:

- The policies studied can bring stabilisation of energy demand and CO_2 emissions *after* 2010 within reach – but no more. Until 2010, they do not show significant effects and help little in the attainment of Kyoto targets.

- Effective policies are available for containing passenger-vehicle energy demand. If measures to hasten fuel-intensity improvements in cars and light trucks continue to be tightened, fuel-demand increases from this mode after 1997 can be contained at low levels until 2020.

- Road freight in small and large trucks accounts for a large share of the increases under the combined scenario. It contributes the most to freight activity and benefits from modal shifts that drive continuous energy-demand growth in freight transport. Rather strong economic and regulatory measures beyond those assumed might be necessary to contain its growth.

- Aviation-fuel demand growth is a major concern. The assumed fuel tax did not have a significant effect. More stringent measures may be justified by the global environmental impacts of aviation that go beyond CO_2 emissions.

- Growth in passenger and freight transport demand remains a long-term problem. It is slowing, but compensating for its effect on energy demand with fuel-intensity improvements alone appears infeasible and insufficient to achieve actual fuel-demand reductions in the longer term.

Toward a Vision of a Sustainable Transport Future

The two transport scenarios in the *World Energy Outlook* indicated with only modest differences steady increases in travel, oil use, and CO_2 emissions in all OECD regions through 2020. How does this compare to a vision of a truly sustainable transportation energy future? While there is no unequivocal view on what a sustainable transport path would look like, it is useful to initiate discussion on this question by describing one possible pathway that *could* lead to a sustainable future. Such a "straw man" can focus the debate on the subject and the types of changes that would have to be made to transport systems to get onto that path.

One major OECD effort, known as the Environmentally Sustainable Transport (EST) effort, has recently provided a very ambitious scenario for transport that is briefly presented below. This exercise has produced the following targets for achieving environmental sustainability in transport:

- Total emissions of CO_2 from transport should not exceed 20 per cent of such emissions in 1990.
- Total emissions of transport-related VOCs and NOx should not exceed 10 per cent of such emissions in 1990.
- Land use due to infrastructure for the movement, maintenance, and storage of transport vehicles should be developed in line with local and regional objectives for air, water and ecosystem protection. Transport activity should use progressively smaller portions of land.

The EST project also provides ambitious targets for noise reduction and particulates, but does not provide specific limits for oil use in transport. All objectives are assumed to be met by the year 2030. Without discussing the merits of any one of these objectives, one can nevertheless identify some of the policy measures and/or technological developments necessary in order to bridge the gap between the current scenario and the EST scenario. Grouped by themes, they are discussed below.

Fuel Economy: Accelerating the Adoption Rate of Advanced Technologies

To achieve early market penetration of advanced technologies such as hybrid electric and fuel cell propulsion systems, governments may need to provide relatively large rebates on these vehicles in order to make them attractive to consumers and producers. As an alternative to a broad feebate programme, more targeted incentives could be used to overcome the very high initial capital costs and risks of producing such vehicles, and encourage consumers to purchase them. Vehicle purchase incentives could be implemented in the form of price subsidies, either for vehicles possessing specific technologies or simply for vehicles meeting a certain standard for fuel economy or CO_2 emissions per mile. Such subsidies could also reflect the level of pollutant emissions, for which the vehicle is certified.

An example of "next-generation" technology price incentives is found in Japan, where there is a price incentive of about US$ 3,500 per vehicle for hybrid-electric vehicles. Japan is the first country with significant sales of these advanced technology vehicles, with over 25,000 hybrids already sold there. A time-phased approach including large initial subsidies per vehicle that decline over time as vehicle sales increase and their costs decline may produce an even more rapid build-up in production volumes and the rate of learning.

Alternative Fuels – Moving to a Near-zero Emission System

There are few alternative fuels that provide dramatic CO_2 reductions relative to gasoline. However, all non-petroleum fuels can provide dramatic reductions in oil imports and thus improve sustainability through a reduction of energy supply risk. In the long term, perhaps the most promising path for virtually eliminating the direct use of petroleum fuels, and thus direct emissions of CO_2 and most other pollutants in transport, is the hydrogen fuel cell. Once all vehicles operate on hydrogen fuels, they will be potentially renewably fuelled (if a renewable source of hydrogen is developed), and will produce water as their only emission. This would eliminate most of the direct environmental impact of cars. Their impacts on land-use would be the primary remaining issue.

As described in the previous section, one transition path is for governments to push hard in the direction of fuel cell vehicles, perhaps initially ones using gasoline fuel to gain experience and production capacity. After 2010, a strong push toward hydrogen fuel cells would be needed, including the development of hydrogen-refuelling infrastructure. Policies such as targeted vehicle incentives could also be applied as tax credits for installation of refuelling infrastructure, or could come in the form of direct government investments. The level of investment needed to pay for infrastructure and the incremental costs of fuel cell vehicles over 30 years, are likely to be massive. Society's willingness to pay these costs will largely reflect the level of determination to move transport onto a more sustainable course.

Reaching an End to Travel Growth

Apart from imposing extremely high and continuously increasing levels of fuel taxation, which may be socially unacceptable, reining in travel can probably only be accomplished through a co-ordinated package of different measures. This package must both discourage personal vehicle travel and encourage use of other modes and alternatives to travel altogether. It would probably need to include many of the following elements:

- Widespread electronic roadway pricing at least on all limited access highways, cordon pricing in urban areas.
- Increased parking costs and better enforcement in many areas.
- Much stricter land-use controls, potentially including growth boundaries, improvements in pedestrian and non-motorised travel infrastructure, restrictions on parking supply in new

construction, strong incentives for mixed use, multimodal developments, and other elements that help end the current practice of building car-dependent communities.

- Incentives to speed the rate of adoption of telecommuting, especially 1-2 day-per-week home based telecommuting.
- Carefully co-ordinated improvements to public transit that include measures to increase average load factors for existing systems. Simply expanding transit service can lead to empty buses, causing increases in CO_2.
- Intensive education and publicity campaigns to encourage a much less car-oriented society. The seeds of such efforts are now being planted in many places, as reflected, for instance, by "car-free" days.

All of these measures are already well known and have been implemented in various forms in many cities around the world. But few authorities have implemented these measures sufficiently to come anywhere near ending the growth in personal vehicle travel. Even in cities with good transit service and good facilities for non-motorised travel significant problems of traffic congestion and air pollution remain.

Increasing the Efficiency of Freight Movement

One of the most intractable policy issues in the transport sector is freight transport. There are, however, at least four ways in which freight transport, energy use, and greenhouse gas emissions could be significantly reduced. While not exhaustive, the following list provides some of the potentially most important areas for government action over the medium-term:

- Increasing investment in rail and water-born transport modes and in intermodal facilities: this reserves truck transport for relatively short trips. Pricing incentives can also play an important role in pushing such shifts to the maximum extent feasible.
- Fuel shifting in medium and heavy trucks: just as hydrogen fuel cell propulsion systems can bring large sustainability benefits for

light-duty vehicles, they can do the same for most types of heavy-duty vehicles. Fuel cell engines are being developed for many types of heavy-duty vehicle, including trucks, buses and trains. Applications in some market niches such as urban transit buses may become cost-effective well before light-duty vehicle applications. Central refuelling of fleets may also make hydrogen fuel cell applications feasible earlier than for light duty fleets.

- Incentives and government infrastructure support to rapidly develop much more sophisticated systems and linkages between suppliers and their clients: this could significantly increase the efficiency of goods delivery in metropolitan areas. Current low levels of truck load, "back-haul" trips where trucks run empty, and unsuited sizes of trucks for different applications need to be addressed and could result in far fewer kilometres per ton-kilometre of goods movement. Systems are being installed in many places to improve freight logistics, but the relative fragmentation of the truck industry impedes the types of co-operation needed to maximise potential efficiency gains.
- Rationalisation of locations of goods producers, distributors, and end-users. Many "upstream" non-retail activities could be moved much closer to other steps in the chain taking a product from production to the end-user. This is a particularly difficult problem to address, however, because shipping costs are typically such a small part of location decisions that they are not likely to cause significant changes. Long-term zoning decisions, however, might be effective.

Conclusions

This chapter has sketched out some of the key issues regarding transport energy sustainability and traced several possible scenarios for transport over the next 20 or 30 years. It is clear that the Reference Scenario path is not even remotely on a sustainable path. Even a policy package featuring some of the more realistic policies that could be applied in the next few years appears likely to fall short of "bending the curves" of oil use and CO_2 in transport. The OECD/EST "vision" of a sustainable transport future in 2030 is currently just that – an internally consistent think-piece that captures the imagination of policymakers but that is not attainable with the policy measures currently under consideration. Nevertheless, some of the ambitious policy options outlined above might be developed into concrete policy proposals over the coming years.

One of the necessary first steps toward a transport sector that is sustainable in all three dimensions is for governments to convince their citizens of the need for aggressive action to alter current unsustainable trends. This includes informing the public about the risks associated with the current path and the importance of transport in reducing oil dependence and CO_2 emissions. Without strong public support and understanding, there will be little political will to enact policies leading to a sustainable path. Currently, even low-cost action that provides clear societal benefits beyond fuel savings and CO_2 reductions can be blocked by small groups of stakeholders who may stand to lose if not effectively compensated.

Governments also need to overcome "traditions" that no longer make sense such as company car policies, high fixed but low variable taxation, or particularly in non-Member countries, cheap fuel. Governments also need to become more entrepreneurial: the existence of good transit or walking and biking facilities does not necessarily translate into high usage. The public has to be convinced that the benefits that these modes offer – lower traffic congestion, cleaner air, and lower CO_2 emissions – are worthwhile. They will be attainable only if people are willing to make changes to their lifestyles.

Public decision-makers need to identify concrete and widely agreed sub-goals of sustainability and sketch out pathways to reach them. Developing such scenarios in co-operation with stakeholders and disseminating them widely may be critical in generating social consensus on concrete sustainability objectives as well as on the distribution of the costs of achieving them.

PREVENTING CLIMATE CHANGE RISK THROUGH FLEXIBLE MARKET MECHANISMS

Climate Change, Energy and Sustainable Development

Fulfilling their commitments to reduce greenhouse gas emissions in order to minimise the risk of climate change is perhaps the most concrete medium-term sustainability challenge for IEA/OECD countries. In many ways, this most tangible of sustainability goals is a bellwether for a commitment to sustainable development in general. Foregoing serious efforts to reach the targets agreed to under the Kyoto Protocol would most likely be interpreted widely as a sign that sustainable development itself has slipped down the priority list on policy agendas.

For developed and developing countries alike, the projected change in climate due to the accumulation of man-made greenhouse gases in the atmosphere will have significant impacts on water resources, food security, ecosystems, health and sea-levels [105]. Uncertainties remain in the science of climate change, especially in regard to its precise regional impacts, but due to the irreversible nature of the likely changes, waiting for definitive answers before taking action is not a sustainable option.

The United Nations Framework Convention on Climate Change adopted in 1992 and its Kyoto Protocol of 1997 take a precautionary approach to these problems. The ultimate objective adopted by Parties under the UNFCCC is to stabilise atmospheric concentrations of greenhouse gases at a level that would prevent dangerous man-made interference with the climate system. This

105 On 13 November 2000, Robert T. Watson, the Chair of the Intergovernmental Panel on Climate Change at the Sixth Conference of Parties addressed the United Nations Framework Convention on Climate Change with these words: "The overwhelming majority of scientific experts, whilst recognising that scientific uncertainties exist, nonetheless believe that human-induced climate change is inevitable. ... The question is not whether climate will change in response to human activities, but rather how much (magnitude), how fast (the rate of change) and where (regional patterns)."

would require roughly halving global emissions of greenhouse gases over the next century.

The extraction, transport and combustion of coal, oil and natural gas accounts for the majority of greenhouse gas emissions to the atmosphere. Fossil fuel combustion creates roughly four-fifths of total greenhouse gas emissions in the form of CO_2. The energy sector also contributes to the majority of CH_4 emissions through coal, oil and gas extraction and transport, and to N_2O emissions through energy and transport[106]. During the period 1997-2020, based on a stable policy environment, global energy-related CO_2 emissions are projected to grow by 60 per cent in the Reference Scenario of the IEA's *World Energy Outlook 2000*. This projected increase presents a particular challenge to the countries listed in Annex B of the Kyoto Protocol, which have committed themselves to *reduce* their emissions by the budget period 2008-2012. Except for transition economies, very few of these countries are in fact likely to meet their goals without additional action (see the table below).

Table 10

CO_2 Emissions from Fossil Fuel Combustion and Kyoto Targets[107] (Million Tonnes of CO_2)

	1990	1998	1998/90	Target
Annex I	**13 825.8**	**13 383.4**	**–3.2%**	**x**
Annex II	**9 957.**	**10 791.9**	**8.4%**	**x**
North America	5 265.2	5 887.0	11.8%	x
Canada	421.3	477.3	13.3%	–6%
United States	4 843.8	5 409.8	11.7%	–7%
Europe	3 361.1	3 435.3	2.2%	x
Austria	58.8	61.6	4.9%	–13%
Belgium	106.2	122.5	15.3%	–7.5%
Denmark	51.0	57.3	12.3%	–21%
Finland	53.4	59.7	11.9%	0%
France	368.6	375.5	1.9%	0%

106. IEA (2000), World Energy Outlook *2000.*

107. IEA (2000), CO_2 Emissions from Fossil Fuel Combustion.

	1990	1998	1998/90	Target
Germany	967.2	857.1	–11.4%	–21%
Greece	69.9	82.6	18.3%	+25%
Iceland	2.0	2.1	5.5%	+10%
Ireland	32.2	38.4	19.4%	+13%
Italy	401.6	426.0	6.1%	–6.5%
Luxembourg	10.5	7.2	–31.3%	–28%
Netherlands	156.8	171.4	9.3%	–6%
Norway	28.5	34.3	20.6%	+1%
Portugal	39.9	54.3	36.1%	+27%
Spain	211.6	254.0	20.1%	+15%
Sweden	51.8	53.5	3.4%	+4%
Switzerland	41.1	40.8	–0.8%	–8%
Turkey	137.8	187.5	36.0%	none
United Kingdom	572.3	549.5	–4.0%	–12.5%
	1 331.2	*1 469.6*	*10.4%*	*x*
Australia	258.7	310.7	20.1%	+8%
Japan	1 048.5	1 128.3	7.6%	–6%
New Zealand	24.1	30.5	26.9%	0%
Economies in Transit.	**3 868.4**	**2 591.5**	**–33.0%**	**x**
Belarus	..	65.3	..	none
Bulgaria	71.3	48.6	–31.8%	–8%
Croatia	..	18.8	..	–5%
Czech Republic	150.4	120.8	–19.7%	–8%
Estonia	..	15.4	..	–8%
Hungary	67.6	57.4	–15.1%	–6%
Latvia	..	7.9	..	–8%
Lithuania	..	15.6	..	–8%
Poland	348.5	320.2	–8.1%	–6%
Romania	166.6	94.6	–43.2%	–8%
Russia	..	1 415.8	..	0%
Slovak Republic	54.2	37.3	–31.2%	–8%
Slovenia	12.7	15.3	20.7%	–8%
Ukraine	..	358.8	..	0%
Non-annex I	**6 826.1**	**8 622.2**	**26.3%**	**none**
Africa	598.9	728.7	21.7%	none
Middle East	600.1	924.2	54.0%	none
Non-OECD Europe	119.3	79.2	–33.6%	none
Other Former USSR	565.2	327.8	–42.0%	none
Latin America	922.0	1 222.7	32.6%	none
Asia (excl. China)	1 631.3	2 446.5	50.0%	none
China	2 389.3	2 893.2	21.1%	none
Mar. bunkers	**355.7**	**398.3**	**12.0%**	**x**
Av. bunkers	**282.0**	**322.0**	**14.2%**	**x**
World total	**21 289.6**	**22 725.9**	**6.7%**	**x**
Annex B	**3 573.8**	**13 130.6**	**–3.3%**	**x**

Notes: The targets apply to a basket of six greenhouse gases and take sinks into account. The overall EU Kyoto target is -8 per cent, but the member countries have agreed on a burden-sharing arrangement as shown above. Because of different base years for different countries and gases, a precise "Kyoto target" cannot be calculated for total Annex I or total Annex B countries. Annex I countries are industrialised countries (OECD Members as of 1992 and economies in transition). Annex B countries are those with commitments under the Kyoto Protocol.

It is clear, then, that the energy sector needs to be at the core of action to reduce the threat of global climate change. To facilitate the achievement of the emission reduction targets, efforts are underway to develop mechanisms that will add flexibility to policies for the reduction of greenhouse gas emissions and help to keep the costs of reaching the targets at a minimum. This chapter provides a discussion of the key mechanisms: Emission Trading, Joint Implementation and the Clean Development Mechanism. Because emission trading mechanisms and their effects are complex efforts have been made to model them. The discussion that follows draws upon the results of these modelling analyses.

Emission Trading and Other Flexibility Mechanisms to Reduce the Cost of Compliance

Early on, countries with commitments to reduce greenhouse gas emissions sought ways to co-ordinate their actions in order to reduce the cost of achieving their emission objectives. The co-ordination of domestic policies turned out to be difficult. Once the negotiations focused on legally binding emission objectives, the idea of international trade in greenhouse gas emissions emerged as an acceptable tool to help bring down the cost of greenhouse gas mitigation[108].

Two instruments to reduce pollution at minimum cost to society have been identified: a tax paid on each unit of the targeted pollutant and a system of tradable permits applied to the sources of the pollutant. In an idealised setting, taxes and tradable permits lead to the same optimal outcome: the equilibrium price of tradable permits would be equal to the level of the tax and both would correspond to the marginal damage connected with the

108. The difficulties of domestic policy harmonisation can be illustrated by reflecting on the following hypothetical questions. Could taxes on gasoline and other fossil fuels be harmonised across countries of the OECD and Eastern Europe, when the price of a litre in the United States is roughly a quarter of that paid by European citizens? Could efficiency standards for home insulation be set at the same levels in Canada and Greece?

externality. In both systems, those who pollute eventually face the same cost for their emissions, respond to the same incentive to reduce them, and therefore achieve the same environmental objective at minimum economic cost.

Due to uncertainty about the magnitude and precise nature of climate change impacts, the magnitude of a pollution tax that would do the job is difficult to determine. The question of the discount rate to be used to incorporate future costs and benefits in current decisions adds another layer of complexity. In the end, the emission targets agreed upon at Kyoto were set on the basis of various criteria, including: countries' willingness to pay, per-capita emission levels, projected greenhouse gas emissions and domestic mitigation options. Hence, the Protocol cannot be viewed as striving for an optimal climate change policy, but rather as a political agreement that will be a step toward a more sustainable path.

Trading emission reductions across Annex I countries (and beyond through the Clean Development Mechanism) would be a step toward assuring the sustainability of climate change mitigation options. By helping cut the cost of reducing greenhouse gas emissions, trading should alleviate concerns about the economic implications of reduction actions. An international price on greenhouse gas mitigation would also be a step toward the internalisation of the climate change externality in all human activities.

Flexibility Mechanisms to Reduce Costs of Emission Reductions

Under the Kyoto Protocol, any Annex I Party with an emission objective is allowed to transfer units of emissions to another Party if its 2008-2012 emission levels are lower than its initially-allocated amount (the assigned amount of emissions, or AAUs in the parlance of the Protocol). Two mechanisms have been introduced for such transfers: "Joint Implementation" (JI) is based on emission reductions arising from specific projects; international "Emission Trading" (ET) allows governments to trade on the basis of their country's emission inventory. In addition, the Protocol introduces a "Clean Development Mechanism" (CDM), which authorises the crediting of certified emission reductions achieved through sustainable development projects in developing countries (non-Annex I Parties) to the commitment of Annex I Parties [109].

At the end of the commitment period, a Party is declared in compliance with its emission commitment if its emissions are less than or equal to its assigned amount, adjusted for transactions under the flexibility mechanisms:

> Emissions 2008-2012 = Initial assigned amount under the Protocol + adjustments from land-use, land-use change and forestry + acquired units (through JI, CDM and ET) – transferred units (JI and ET).

These flexibility mechanisms cannot be discussed in abstraction from the underlying emission objectives agreed upon by industrialised countries. On the one hand, most OECD countries have agreed to objectives that imply a significant reduction from their current emission trends. On the other, countries with economies in transition were given some flexibility to account for the recession that has been plaguing their economies since 1989 and led to a sharp decline in their emissions, especially CO_2 (see above). It is expected that under current circumstances the emissions of these countries will remain well under their assigned amounts and that they will therefore have large amounts of emission permits for sale on the international market, without having taken measures specifically aimed at reducing greenhouse gas emissions. These projected reductions available for sale are referred to as "hot air".

A number of rules are necessary to make these mechanisms operational. The main design and regulatory issues involved are addressed in the remainder of this section.

109. Reductions under the Clean Development Mechanism can be counted as they occur, anytime between 2000 and 2012, unlike those transferred under Joint Implementation and Emission Trading, which apply to the 2008-2012 time frame.

Emission Trading: Rules and Debates

Emission trading under the Kyoto Protocol is defined in relation to the following specifications. The emission objectives or "assigned amounts" refer to the emissions of six greenhouse gases (CO_2, CH_4, N_2O, HFCs, PFCs, and SF_6) from the energy sector, industrial processes, solvents and other product use, agriculture, and waste. Adjustments are possible for changes in carbon stocks as a result of afforestation, reforestation and deforestation as well as for other activities in the land-use change and forestry sector. The assigned amounts are expressed as averages over five years (2008 to 2012) which provides some flexibility over time.

Emission trading raises a number of important issues. The first one is the eligibility of participants. Joint Implementation and the Clean Development Mechanism explicitly authorise participation of legal entities other than governments such as private companies, foundations or NGOs in the development of projects, while governments retain the responsibility for compliance with countries' emission targets.

Another important issue is the question of liability. Domestic tradable permit systems, such as the U.S. SO_2 allowances programme under the Clean Air Act, are supported by a strong legal system. Under the Kyoto Protocol, without penalties, there is a risk of governments cheating (for example, selling emission reductions not actually realised) if they see no negative consequence for their behaviour[110]. Misuse of trading could endanger the whole mechanism. An alternative could be to put the liability on the buyer. Buyers would then be encouraged to check the prospects of compliance by the originating Party in order to assess their own ability to use the acquired permits for compliance.

A further issue is whether all transactions should take place on an open marketplace or whether bilateral transactions should be allowed. The worry is that governments engaged in international

110. *See also Baron (1999), "An assessment of liability rules for international greenhouse gas emissions trading".*

transactions would put political considerations ahead of economic ones and would do so more easily in bilateral transactions. At the same time, forcing all transactions to go through the market will probably not provide the flexibility that the political intricacies of the UNFCCC process require[111].

A final point under discussion is the question of "supplementarity". Should Parties first concentrate their efforts on the reduction of domestic emissions, and only acquire emission permits from other Parties as a supplement? Or should the allocation of effort between domestic and international actions be left to the market, so that overall costs are minimised?

On the one hand, if countries defer action through the acquisition of "hot air", it will be more difficult for them to accept deeper cuts in emissions at a later stage. On the other , the price emerging from the trading regime could be a potent tool to encourage policymakers to take action, including domestically. There is not a clear-cut answer to this question. One can only note that proposals to implement supplementarity have an important implication – restricting demand would reduce the price of traded emission permits, to the benefit of buyers[112]. Restricting supply – how much transition economies can sell – would remove any incentive for these countries to undertake further reductions.

The Clean Development Mechanism: Rules and its Contribution to Sustainable Development

The Clean Development Mechanism (CDM) aims at generating projects in developing countries to reduce greenhouse gas emissions and contribute to sustainable development in other ways. Once certified, reductions can be credited. In order for the CDM to become operational, Parties must agree on its institutional structure and on guidelines to determine the emission reduction from a

111. *See Baron (1999), "Market power and market access in international greenhouse gas emission trading" and (2000), "Market access issues in international GHG emission trading".*

112. *EC (2000),* Climate Negotiation and Emission Trading: Economic Insights from European Models.

specific project. Contrary to emission trading, every project must be assessed independently before reductions can be certified because developing countries have not adopted quantified objectives.

The key question in relation to any CDM project is: what would emissions have been without the project? This is referred to as the environmental "additionality" question. It is complicated by the fact that both the investor and the host party would benefit from inflating the quantity of reductions achieved by specific projects. This is why the question of "baselines" has drawn so much attention.

There are different ways to set up an emissions baseline [113]. Ideally, baselines should be credible, transparent, simple and inexpensive to establish. In practice, drawing up baselines is likely to involve trade-offs among these criteria.

The calculation of the emission benefits of most projects so far undertaken has been based on project-specific baselines [114]. Work on multi-project baselines is intended to standardise emission levels or rates. The approach used to determine an emissions baseline for a CDM-project has consequences for the project's transaction cost, transparency and administrative feasibility (including data, monitoring and reporting requirements), as well as for its environmental additionality. Data, monitoring and reporting requirements are important because they affect the costs and administrative feasibility of project preparation and review. Project-specific baselines have relatively heavy data requirements and may require some monitoring of current activities before the actual project or activity starts. Using multi-project baselines requires less monitoring of the pre-project situation [115].

---

113. The discussion of baselines for Clean Development Mechanism draws from OECD/IEA (2000), Kyoto Mechanisms, Monitoring and Compliance – A selection of recent OECD and IEA analyses on the Kyoto Protocol *and OECD/IEA (2000),* Emission Baselines – Estimating the Unknown.

114. These projects were undertaken under an earlier designation – "Activities Implemented Jointly" – a mechanism introduced to explore the feasibility of project-based trading mechanisms. The Clean Development Mechanism is likely to replace the earlier mechanism.

115. The cost of establishing an emissions baseline is one component of the transaction costs associated with JI- and CDM-projects. It should ideally be kept as low as possible to encourage investment through these mechanisms.

The transparency of a baseline also varies with different baseline approaches. In general, the more project-specific is a baseline, the more documentation will be needed to make it transparent. The environmental additionality of a CDM-project can be affected by the baseline approach, as it can influence the potential level of doubtful emission reductions that would artificially inflate the number of credits resulting from a project. The environmental additionality of individual projects is also clearly correlated to the level of stringency of the baseline that varies with the assumptions used in setting up the baseline. For example, if both India and Brazil use the same multi-project baseline approach for new electricity projects at the level of their current average emissions, gas-fired electricity projects could generate certified emission reductions if they were undertaken in India, but not in Brazil.

This discussion suggests that it would be desirable to minimise baseline complexity, as long as the ability to determine "what would have happened otherwise" is not compromised. It also suggests the need to make trade-offs on baseline stringency with a view to the overall global environmental effectiveness of project-based mechanisms. An optimal strategy needs to take into account also the need for a high volume of projects. A large number of adequate projects could be more beneficial for the environment than a small number of superior projects.

How Much can Emission Trading Reduce the Cost of Meeting Reduction Goals?

Insights from Modelling

There has been a great deal of research on the cost savings that can be expected from the inclusion of emission trading in the Kyoto Protocol. Alternative scenarios have been produced in global models that enable a comparison between autarchic reductions (i.e., if each country or region meets its obligation domestically) and the international emission-trading regime allowed by the

Protocol. Because economic theory predicts that trading benefits both buyers and sellers, these have typically been exercises in estimating the magnitudes of trading gains, rather than demonstrations that gains will be forthcoming in the first place. In addition, the goals set at Kyoto leave a comfortable margin for increases in emissions by most countries with economies in transition. The transfer of the resulting unused emission allowances would clearly benefit both the seller – in the form of a windfall profit from an economic recession – and the buyer, for whom reducing emissions domestically would be more costly.

The IEA has constructed an emission-trading scenario among Annex B Parties under the Protocol. The scenario is based on the World Energy Model, which in turn is based on econometric estimates of the links between detailed economic activities, energy prices and energy consumption, as well as a linear-programming module with several technologies for power generation and a module for fossil fuel supply prospects. The analysis considered CO_2 from fossil fuel combustion for five groups of countries with objectives for emission reductions: North America, OECD Pacific (excluding South Korea), OECD Europe (excluding Turkey), Russia, and other countries with economies in transition (including Ukraine).

Because an energy model was used to develop the scenario, the only greenhouse gas emission considered was CO_2. The agreed emission reductions were expressed in percentage terms as applied to CO_2 emissions. The *World Energy Outlook* projects a considerable gap between the Kyoto commitments and projected CO_2 emissions by 2010 for the three OECD regions. In contrast, Russia and Ukraine/Eastern Europe will have emissions in 2010 much lower than their Kyoto commitments. Total CO_2 emissions evolve differently in each of the three OECD regions – their increase will be greatest in North America, where they are expected to exceed the combined commitment by 38 per cent, but less in OECD Pacific (24 per cent) and still less in OECD Europe

(17 per cent). The two non-OECD regions, Russia and Ukraine/ Eastern Europe would have CO_2 emissions in 2010 below their commitments by 39 and 35 per cent respectively, thus creating a reservoir of "hot air". The combined emissions of Annex B countries are projected to lie 11 per cent above the combined commitment.

The elaboration of the trading scenario is based on the above projections and on marginal abatement cost curves for each region (Figure 23). The marginal abatement cost curves were calculated by imposing successively rising carbon-tax rates on the World Energy Model. At each rate, the model yielded different carbon-dioxide emissions lower than the Reference Scenario. Thus each carbon-tax rate corresponded to a certain gap between emissions with taxes and emissions without taxes. To derive a marginal abatement cost curve, these gaps were taken as abated emissions corresponding to the different carbon tax rates, identified as implicit cost figures.

Figure 23

Marginal Abatement Cost Curves for the Five Trading Regions

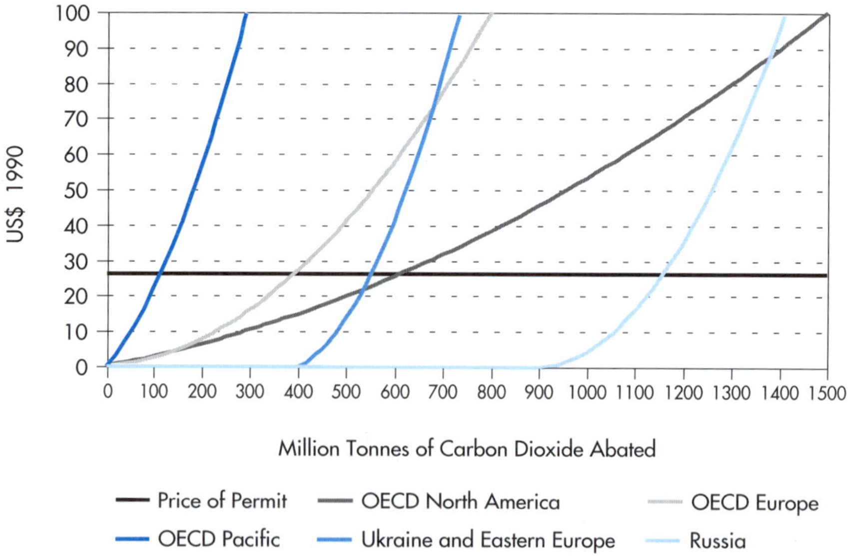

The market-clearing price under an emission-trading scenario is derived from these marginal cost curves and the overall gap between the collective target- and projected-emissions. It is equal to the marginal cost for which all five regions meet their collective emission reduction objective. The curves then indicate traded quantities region by region, as reported in the table below.

The trading price in the progressive action base case is US$ 26 per tonne of CO_2 (US$ 32 in today's dollars) or US$ 95 per tonne of carbon. Based on these figures, trading emission reductions among developed countries would considerably reduce the cost of compliance with the Kyoto objectives, by 63 per cent for North America, 55 per cent for the Pacific region, and 29 per cent for Europe. For the three OECD regions together, the cost savings would reach 58 per cent from a "no-trading" scenario. This corresponds to 0.3 per cent of their combined gross domestic product (GDP). Russia and Ukraine/Eastern Europe would gain 5.9 and 4.6 per cent of GDP, respectively, from participating in emission trading. Clearly, economic sustainability would benefit from emission trading.

This scenario does not include the possibility of a second commitment period following immediately after the Kyoto targets: these projections may be altered if countries must meet more stringent reduction levels from 2013 onward, as they would incorporate the cost of such targets in their first-period trading decisions. The results on the sellers' side of the market are also striking, with emissions at 50 per cent of what they were in 1990 as more reductions are achieved to meet the demand for emission permits from the OECD regions.

Reality Check: an Experimental Market for Emission Trading

The efficiency of tradable permits is well demonstrated theoretically and supported by economic models. However, its implementation in the framework of the Kyoto Protocol and other real-world considerations raises a number of questions. For

Table 11

Carbon Dioxide Emission Trading – Progressive Action (Million Tonnes of CO_2 and Million 1990 U.S. Dollars per Year from 2008 to 2012)

Trading Price for a Tonne of CO_2: US$ 26						
	2010 Reduction Target	Traded Quantities (Imports +, Exports –)	Domestic Abatement	Average Annual Cost of Commit-ment with Trading	Average Annual Trading Cost as % GDP	Average Annual Benefits of Trading as % of GDP[1]
North America	1,882	1,274 (68% of target)	608	39,842	0.36	0.61
Europe	631	240 (38% of target)	391	9,831	0.1	0.04
Pacific	318	204 (64% of target)	114	6,593	0.14	0.17
Russia	-908	-1,166 (hot air 78%)	258	-27,925	-5.87	5.87
Ukraine & EE	-401	-552 (hot air 73%)	151	-12,761	-4.62	4.62
Total: Gross (net)[2]	2,831 (1,522)	1,718 (61% of total)	1,113 (1,522)	56,266 (15,579)	0.22 (0.06)	0.31 (0.49)

Notes: 1. The average annual benefit from trading indicates the difference between the costs of fulfilling the Kyoto commitments without trading and the costs with trading during the budget period. Given that carbon permits would be internationally traded commodities, the underlying GDP figures have been calculated on the basis of U.S. dollars converted at real exchange rates. 2. The "gross" numbers indicate the sums for the three OECD regions. The "net" numbers indicate the sums of the respective magnitudes for all Annex B regions.

instance, models assume that all emission sources would participate in emissions trading, or that they could be included by way of a "blanket" policy, such as a uniform tax on greenhouse gases or a fully comprehensive domestic emission-trading regime. In the case of a tax, it would need to be constantly adjusted to reflect the international trading permit price, which would be problematic.

In the case of domestic trading systems, options are available in theory to include all of a country's sources (for example, by way of a quota allocation to importers and suppliers of fossil fuels), but this is not a probable policy scenario for most countries participating in an international system. The European Commission evaluates at 45 per cent the share of large-scale energy-intensive emitters that could practically be included in an EU-wide system [116]. Other sources and sinks, mostly because they are too small to be monitored in a cost-effective fashion, would need to be covered by specific policies and measures, and may not therefore have access to the international emission-trading regime. Greenhouse gas mitigation policies in the real world will probably not resemble the comprehensive emission trading or uniform carbon tax regimes assumed by economic models.

In this situation, experimental economics can prove useful as a way to test how markets would fare in conditions that are closer to the real world constraints of policymaking. The IEA organised a simulation exercise for that purpose [117]. It involved 16 delegations from market and transition economies, as well as a few private-sector stakeholders. The purpose was to learn about the development of international trading under the conditions set by the Kyoto Protocol.

Simulating Domestic and International Climate Change Policy

The 24 participants were asked to elaborate strategies to reduce emissions domestically and to trade CO_2 if that made economic sense. They used individual country or sector models for that purpose. These included a business-as-usual projection of CO_2

116. *EC (2000),* Green Paper on Greenhouse Gas Emissions Trading within the European Union.
117. *Baron (2000), "International Emission Trading – A Real-time Simulation".*

emissions for 2000-2012 and a policy instrument equivalent to a tax on carbon emissions. The tax could be introduced and adjusted at any time to reduce emissions toward the goal; however, it only affected emissions in the next year, to reflect the lead-time between the decision and implementation of any policy. Future emission levels were not known with full certainty, consistent with what would occur in the real world[118].

The individual models were to be used to determine the marginal costs of reductions under various strategies, which would enter into determining the participant's allocation of effort between trading and domestic mitigation. All offers made on the exchange were anonymous. Participants could not identify the country selling or buying units; price was the only discriminating factor. In addition to using the public exchange, participants could enter into bilateral transactions, in which case the identities of buyers and sellers were known.

Through price competition a relatively stable price emerged quickly in the virtual currency that was chosen for the simulation, after an initial stage of volatility. Participants started trading early; 60 per cent of all traded tonnes were exchanged before 2010. Out of the 2.9 billion tonnes of carbon traded during the simulation, the net transfers among participants were only 2.1 billion tonnes; 800 million tonnes were therefore traded for the purpose of speculation or hedging, which are common features of markets.

Participants needed to formulate domestic emission reduction strategies early so as to avoid drastic and costly adjustments in the future. For that reason, early policy decisions taken by participants were not adjusted to the international price for emission permits. However, since the trading price varied significantly in the early years, such strategies were difficult to implement.

118. The internet-based market was developed by the Laboratory for Experimental Economics and Political Science of the California Institute of Technology (Pasadena, CA). The exchange can be accessed at http://eeps2.caltech.edu/market-carbon. Country #21 and the password ybq will open the account, where all transactions have been recorded.

Two main factors explain price variations observed during the simulation. Uncertainty about emission trends in the starting period and lack of information on the costs of mitigation explain the price volatility observed in the early stage of the simulation. Moreover, near the end of the commitment period domestic actions to bring about additional reductions became increasingly costly. This intensified competition on the buying side and led to a progressive increase in emission permit prices from 2009 onward. In the end, all participating "countries" and "companies" eventually complied with their emission objectives, and did so more cheaply than would have been the case with exclusively domestic efforts.

An important question is the extent to which participants rely on trading to achieve their emission reduction commitments. Figure 24 details the relative share of domestic reductions and acquired emission permits for each participant.

Figure 24

How Did Countries Meet their Emission Objectives?

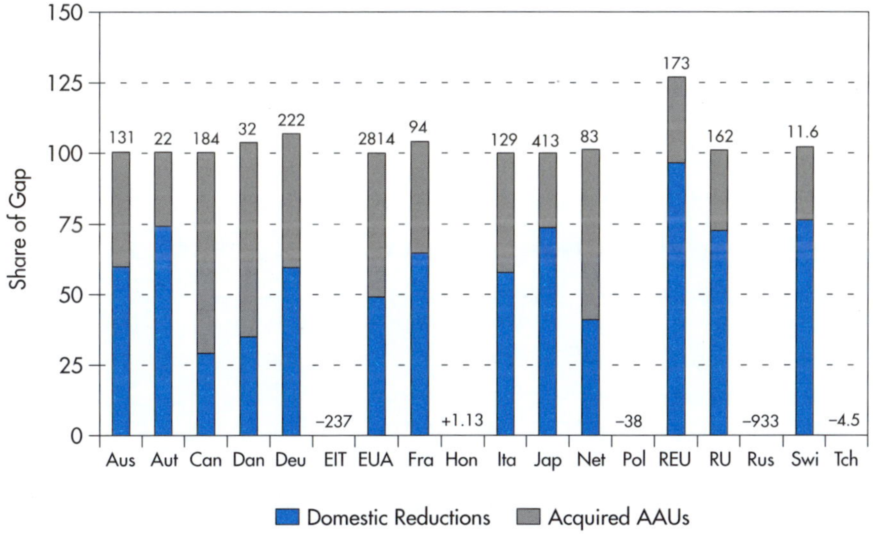

Note: Each participant's total effort is indicated at the top of the bar, e.g., AUS needed to reduce its emissions by 131 MtC in order to meet its emission goal. Net sellers are not represented here.

Another important question is how much participants were able to reduce their costs of compliance through trading. Figure 25 shows the savings achieved by participants, when their strategy in the simulation is compared with the total cost of a purely domestic mitigation scenario. The biggest savings were made by those countries with the largest gaps between domestic costs and the market price – which depended on the initial sizes of the gaps between projected emissions and target levels, and of the cost curves for domestic reductions.

Figure 25

Cost Savings Achieved through Trading in the IEA Simulation

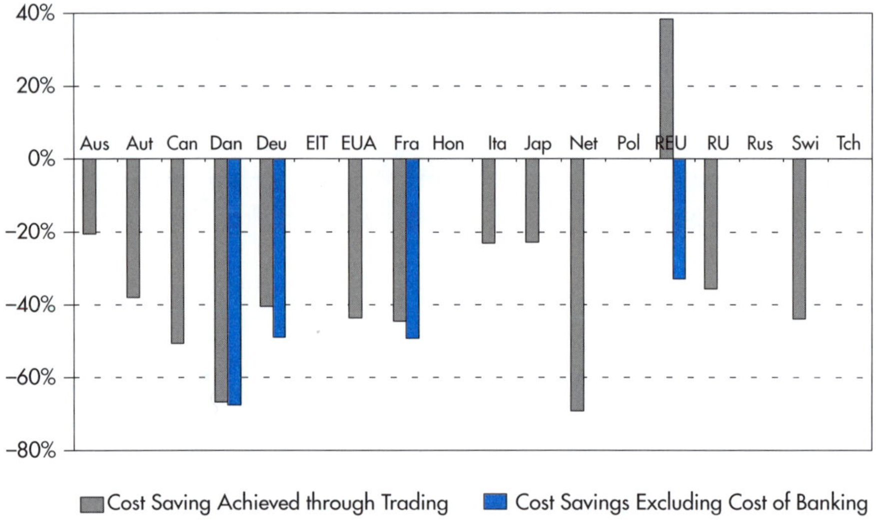

Note: Cost savings are indicated with and without taking into account the cost of banked units at the end of the commitment period.

Despite the large economic benefits that trading brought in the simulation one curiosity remained. Participants did not systematically adjust their assumptions on domestic costs to take account of the international price of traded emission permits. This was a surprise, as doing so would constitute the least-cost strategy.

Several features of this simulation help explain why marginal costs were not systematically adjusted. The explanations offer insights into the difference between theoretical prediction and practical realisation:

- Parties need to start taking domestic measures (and in some cases have already done so) without any certainty about future emission permit prices.
- These measures are unlikely to be fully adjustable to the international price of emission permits. Policy stability may be perceived as more desirable in the long run than taking account of near-term efficiency.
- Future emission levels will remain uncertain from one year to the next because fluctuations in economic growth, energy prices and climate can significantly affect emissions. Such uncertainty calls for a cautious approach by governments and companies alike.

Insights from the Simulation: What Parties Can Expect from Emission Trading

Due to marginal costs not being adjusted, the simulation did not achieve reductions at the very lowest possible cost. As a result of uncertainties and rigidities, the average price observed in the simulation was some 30 per cent above the price at which participants could have met their collective and individual objectives. Note, however, that this high price was not the result of price manipulation by participants.

The simulation raises some key issues. Very few participants set aside emission permits for use in a future commitment period – an option available in the Protocol known as "banking". Participants might have banked emission permits if the emission objectives for the second commitment period had been known and if the cost of compliance had been expected to climb in the future. With full information, the price of emission permits before 2012 would necessarily have reflected this future constraint.

It currently takes about two years to gather data on national inventories. If this remains true in 2008-2012, trading for the first commitment period will extend well beyond 2012, at which point new policies would have no effect on the amount of emission permits available for the commitment period. The simulation optimistically assumed a one-year delay in knowledge of inventories, yet prices still fluctuated from one year to the next as a result of changes in expected emission levels.

A number of conclusions can be drawn from the simulation:

- Emissions trading can work to help cut the cost of meeting the Kyoto Protocol goals, even in conditions that are closer to reality than the perfect setting proposed by macro-economic models.
- This remains true even with the likely policy inertia at country level and the price uncertainty that can be expected from the system.
- Trading by private companies could help countries adjust their efforts to the international price of traded emissions: the more sources that have access to the international emission trading market, the more a country could adjust its domestic cost to the international price and reduce its cost of compliance.
- An emission-trading market would encourage further emission reductions in countries with a low cost of abatement. The question is whether the emissions trading market will be enough to trigger the ambitious policies needed in countries in transition if they are to sell additional emission permits.
- Timely inventories and trading reports are essential to market stability and predictability.
- An early decision on emission constraints after 2008-2012 will provide critical information for the development of the market in the first period.
- International emission trading could accommodate a variety of domestic policy choices.

From Economic Sustainability to Sustainability in all Dimensions

Flexible market instruments could greatly reduce the cost of the first step toward a more stable climate if they are properly implemented, and thus could be a source of improved environmental and economic sustainability. An international emission trading mechanism could also provide valuable information on the cost of achieving the Kyoto Protocol's emission objectives. And to the extent that emission reductions could provide an exportable item for developing countries (for example, in the context of the Clean Development Mechanism), the mechanisms might also contribute to social sustainability. The number of countries that have implemented or are actively studying tradable permit regimes for greenhouse gas mitigation and green power is a sign that these instruments are gaining support, in preference to the policy alternative, carbon taxes [119]. At the same time, there is still very much a need to think in terms of policy packages and policy integration, and not in terms of exclusive policy choices.

First, the flexibility mechanisms discussed under the Kyoto Protocol are not "one-size-fits-all" instruments in relation to domestic climate change policy; there is still room for taxation and other regulatory approaches. The latter are especially relevant when market failures cannot be dealt with by way of economic instruments. Tradable permits also would be difficult to apply to emissions from vehicles and to the myriad of small economic entities in the services or residential sectors. And all too often, taxes or absolute emission caps – a prerequisite to establish a trading regime compatible with international emission trading – remain politically difficult. In sum, a package of policies and measures will be necessary to address the variety of sources and sectors responsible for greenhouse gas emissions.

119. *IEA (2000),* Dealing with Climate Change – Policies and Measures in IEA Member Countries.

Second, while climate change is gaining momentum on the political agenda, other environmental concerns should not be set aside when thinking about how best to organise participation in the international emission-trading regime. Local concerns should be integrated with policies to address climate change. When the time comes to apply greenhouse gas emission trading to domestic entities, great care will be needed to maintain a proper balance between the economics of greenhouse gas mitigation and local environmental goals. For instance, if regulations on other emissions are alleviated in conjunction with allowing private participants to acquire emission permits, this is likely to reinforce the popular misconception that tradable permits are rights to pollute. Sustainable environmental policy calls for a broader approach than one strictly focused on greenhouse gases if it is to meet expectations about a sustainable future for the energy sector.

NON-MEMBER COUNTRIES: SPECIFIC CHALLENGES AND GLOBAL IMPLICATIONS

The Growing Weight of non-Member Countries

Globalising energy markets create increasing interdependencies and common vulnerabilities for sustainable development in the energy sector. The growing importance of non-Member countries creates serious challenges for global supply security and greenhouse gas emissions. The fact that worldwide nearly 2 billion people lack access to electricity requires enormous investments to enable them to benefit from basic energy services such as lighting or refrigeration. Technical progress holds great potential to advance non-Member countries in several dimensions of sustainable development, but it is unlikely to be realised without the active involvement of IEA/OECD countries.

It is now well established that the weight of IEA/OECD countries in the global energy balance is decreasing and that of non-Member countries – broadly speaking, economies in transition and developing countries – is increasing. IEA/OECD countries today account for less than 20 per cent of the world's population. They enjoy 80 per cent of global riches (60 per cent in purchasing power parity terms) and account for 53 per cent of the world's annual energy consumption. On average, per capita energy consumption in non-Member countries is only one-fifth of that of IEA/OECD countries. An average citizen of India consumes only $1/16^{th}$ of the energy consumed by each American.

China and India each has more than one billion people – the size of the whole OECD population. Africa has around 750 million people and Latin America and the rest of Asia around 500 million each. Each individual in these developing regions aspires to a better life, which often means heated or air-conditioned homes, motorised mobility or more access to durable consumer goods. Energy is an essential element for meeting those aspirations. Few people would wish to deny every citizen of the earth those aspirations and his or her efforts to meet them. On the contrary, there have been significant efforts by the IEA/OECD countries through their aid programmes to help achieve those aspirations. However, the manner in which developing countries achieve their aspirations in the medium to longer term bears heavily on the sustainability of the whole world, including that of the IEA/OECD countries.

The bulk of the projected future increase in world energy demand will come from the regions outside the OECD area. The IEA projects in its *World Energy Outlook 2000* that the share of OECD countries in the global energy balance will decline from the current 54 per cent to 44 per cent by 2020 while that of developing countries will rise to 45 per cent from the current 34 per cent. The share of transition economies will decrease slightly from today's 12 per cent to 11 per cent in 2020. Developing countries will account for two-thirds of the total increase in world energy demand between 1997 and 2020. Their total energy demand will more than double. The increase in China alone will be equivalent to the increase in all OECD countries. Factors driving this strong growth of energy demand in the developing countries include rapid economic growth, industrial expansion, urbanisation, substitution of commercial for non-commercial fuels, and in many cases low energy prices as a result of subsidies.

As a direct consequence of their fast-growing energy demand, developing countries will also contribute heavily to the increase in the world's CO_2 emissions. In 1997, IEA/OECD countries were responsible for 51 per cent, transition economies for 11 per cent and developing countries for 38 per cent of energy-related CO_2 emissions. By 2020, developing countries are projected to overtake the OECD in prospective shares: 40 per cent for the IEA/OECD countries, 50 per cent for the developing countries and the remaining 10 per cent for the transition economies. Developing countries will contribute over two-thirds (70 per cent) of the incremental growth in global CO_2 emissions between now and 2020. However, even with this strong relative increase, OECD per capita emissions will continue to be substantially higher than those of developing countries.

Figure 26

Increase in Energy Demand by Region, 1997-2020

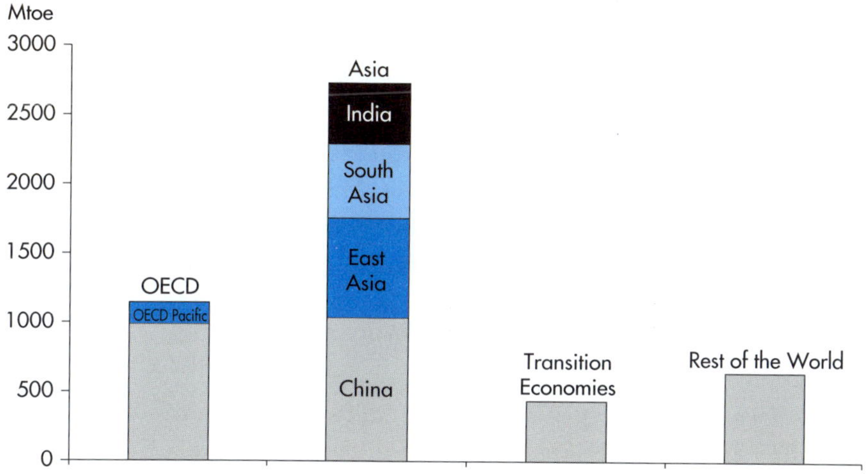

Source: IEA (2000), World Energy Outlook 2000.

Common Sustainability Objectives

These developments demonstrate the importance of engaging non-Member countries in discussions of IEA/OECD countries on sustainable development. Non-Member countries significantly influence all three major dimensions of sustainable development. This becomes evident when each of the issues discussed in the seven previous chapters is considered in relation to non-Member countries. The policy areas presented are in some respects even more crucial for non-Member countries than for IEA/OECD Member countries themselves.

- *Energy supply security* poses acute challenges for developing countries, as they will rely more and more on international markets to meet their galloping energy demand. Most of them, however, do not have emergency oil stocks. They will be hit harder by fuel price hikes, as their economies tend to be more oil-intensive than those of OECD countries. They have lower financial means to pay for the high oil prices that richer economies can afford. Most developing countries also have fewer practical options for energy supply diversity; some options such as nuclear power and LNG are technically unfeasible. Furthermore, most non-Member countries do not have enough political and diplomatic clout to help allay the consequences of an international crisis that may entail supply problems. Energy security is, therefore, becoming an increasingly important preoccupation for many non-Member countries. It is worth noting that some countries like China, Brazil and several ASEAN countries plan to build up emergency oil stocks.

- *Energy market reform* is a pressing issue in many non-Member countries, since it is a pre-condition to attracting capital investment in energy infrastructure. In contrast to IEA/OECD countries, where energy reform is mainly seen as a stimulus for economic efficiency through competition, energy market reform in non-Member countries focuses more on the creation of conditions that would allow the delivery of energy services in a financially sustainable manner. Many non-Member countries also need to succeed in attracting international capital, which tends to flow to more rewarding markets with lower risks.

- *Improving energy efficiency* constitutes a vital challenge, as energy efficiency in most non-Member countries is much below the OECD average. Factors that contribute to this lower efficiency include lower levels of technology; a lack of necessary institutional means, skills and management capacity; economic structures that favour energy intensive industries as a result of the international division of labour; and lower energy prices frequently due to subsidisation. Although most countries recognise the benefits of energy efficiency improvement –

including reduced need to expand supply, improving the environment as well as financial benefits as a result of energy savings and reduced energy imports or increased energy exports – many countries still have difficulties turning policy intentions into concrete actions. Energy efficiency measures tend to be more difficult to finance than energy supply projects. They often require governments to revisit existing policies and to take new initiatives.

- *Renewable energies* and expanding access to energy and energy technologies involve special issues that are discussed below.
- Policies for *sustainable transport* are highly relevant to developing countries because many of their cities are beset by notorious congestion problems. Without appropriate strong actions, the problem can only worsen in the future, as transportation in many countries, including the largest, such as China, India and Brazil, is expected to grow at a vertiginous pace. These countries must adopt effective alternative traffic systems if they are to cope with booming urban development.
- While *climate change risk* is frequently not a policy priority in non-Member countries, it could in fact cause more serious damage to developing countries than to IEA/OECD countries. Indeed, countries in the developing world have lower technological and financial means with which to adapt to the changing, erratic global climate patterns or to prevent the results of natural disasters. Severe droughts, storms, floods and other extreme weather events cause more human casualties when they occur in a developing country. Extreme weather events, as well as the advancing desert across West Africa and the rising waters threatening coastal and island nations, are stark signs of the environmental crises facing developing countries as a result of global climate change. Developing nations therefore have much to gain by joining the developed world's efforts to reduce the emissions of greenhouse gases, despite the fact that their *per capita* emissions are much lower than those of the industrialised world.

The Heterogeneity of Non-Member Countries and the Diversity of Challenges

Unlike the OECD countries, which have comparable levels of economic development, industrialisation, rules of law and market-based mechanisms, the non-Member group is far more heterogeneous. Economies in transition are also clearly distinct from the developing world because of their high degree of industrialisation (see also the box below on Russia). Significant differences also exist in cultures, lifestyles, social and institutional structures, as well as patterns of economic development.

For instance, for China and India, the two largest developing countries, the first challenge in the energy sector is to provide adequate supply to meet the growing energy demand of their people and economies. No less a top priority is the need for China and India to manage their serious environmental pollution that results from energy production and consumption. As both countries use coal as the predominant energy source, one common long-term sustainable development priority for China and India is to diversify away from coal, by developing natural gas, hydropower and other renewable or non-fossil energy resources. Despite these commonalties in their sustainable energy challenges, China and India differ significantly in social and institutional settings and therefore have different priorities in responding to these challenges. For example, energy pricing reform is much more advanced in China than in India; it is therefore a higher priority for India than for China. On the other hand, establishing and respecting the rules of law in markets represents more of a challenge for China than for India.

The resource-rich Middle East countries have different sets of economic and social conditions than the rest of the developing world. Middle East oil producers need to diversify away from their single-commodity-based economies. Opening of the upstream sector to foreign investment is also an important challenge, since the call on that region's oil production is increasing. Foreign investment can also help countries in the region to tap important natural gas reserves and to develop associated processing and export facilities.

African countries are highly diverse. They include a number of OPEC oil producers, such as Nigeria, Algeria and Libya, as well as coal-rich South Africa. Challenges for these energy-rich countries differ greatly from those which are faced by the energy-poor countries of the sub-Saharan region, where desertification and deforestation caused by fuelwood use mutually re-enforce each other.

While developing countries thus share a number of global sustainability problems already familiar to IEA/OECD countries, several sustainability issues are specific to them, such as the use of biomass energy, access to commercial energy services, framework conditions for investment, energy market reform, technology transfer and the often difficult trade-offs between economic development and environmental performance. These issues are addressed in the sections that follow. The discussion draws frequently on examples from those important non-Member countries with which the IEA maintains particularly close relationships – China, Russia, India and Brazil.

Sustainability Issues in Developing Countries

The Use of Biomass Energy

Worldwide, biomass energy (fuel-wood, charcoal, agricultural residuals, etc.) constitutes over 10 per cent of global total primary energy supply. It represents by far the largest renewable energy use in the world, especially in rural areas of developing countries. In China and India respectively, biomass energy constitutes respectively 20 per cent and 42 per cent of total primary energy supply. In many countries, biomass provides the basic energy needs of households, such as cooking and heating.

Box 12

Key Issues for Sustainable Energy Development in Russia

Sustainable energy development in Russia is subject to uncertainty. Lack of confidence regarding the evolution of economic growth and the investment levels required to meet expected growth in energy demand are key areas of concern for Russian energy policymakers. Two additional uncertainties concern the progress of energy sector reforms, which have been lacklustre over the past decade, and the ability of Russia to remain a major energy exporter to Western Europe and the countries of Central and Eastern Europe.

A lack of investment, due to the unstable investment environment in Russia, has led to low reserve replacement and limited turnover in the capital stock, which only exacerbated this problem. Energy intensity in Russia is more than three times larger than in OECD Europe, a legacy of low energy prices and limited payments for energy services.

Several key objectives and actions will help to determine Russia's progress toward a more sustainable energy sector:

- *Sustaining the economic recovery fuelled mainly by external forces (current higher oil prices and the impact of the 1998 rouble devaluation). This will depend on the government's ability to follow through on its talk of significant legal, fiscal and price reforms.*
- *Tapping the enormous potential for improvements in energy efficiency. This will require pricing reform as well as a number of strong regulatory actions by public authorities.*
- *Increasing investment and enhancing the attractiveness of energy sector investment through the introduction of stable investment frameworks, such as Production Sharing Agreements.*
- *Increasing the output of the coal and the nuclear sector to match the planned decrease in the share of gas in Russia's total primary energy supply within next two decades. The most economic option might well continue to be gas even if its current subsidisation ceases.*
- *Given Russia's status as an Annex B country, the pace of reform is also very important for its future level of CO_2 emissions. Joint Implementation projects could help to enhance the stability and attractiveness of energy efficiency investments and help Russia's sustainable development during a time of increased economic activity and energy demand.*

While the use of biomass in IEA/OECD countries is promoted as a positive contribution to sustainable development, it is associated in non-Member countries mainly with negative impacts on all three dimensions of sustainable development. These impacts involve deforestation, local pollution in the form of particulate emissions, and health problems due to cooking in poorly ventilated premises. Children's schooling time is curtailed by fuel-wood and dung collection. Women, particularly in Sub-Saharan Africa and South Asia, spend considerable time collecting fuel-wood at significant cost in terms of foregone production of other goods and services.

Biomass is often considered to be a renewable source of energy. However, unlike other renewable energy sources, it is not always "renewed" because it is harvested without replacement. Indeed, in many parts of the developing world, fuel wood has become more and more scarce. Intensive use of biomass fuels also increases soil erosion or deprives the soil of recycled nutrients. Combustion is often incomplete leading to high emissions of gases, such as carbon monoxide (CO) and methane (CH_4) as well as soot and particulates. Women and children in rural areas and urban slums bear the heaviest burden of these emissions in terms of acute respiratory infections, chronic obtrusive lung diseases, eye problems and low birth weights.

Many studies have shown a strong positive correlation between indoor air pollution and morbidity[120]. In India, acute respiratory infection is the main cause of death (13 per cent). Pneumonia in young children causes more than 300,000 deaths each year out of a total of 500,000. In China, indoor air pollution is estimated to cause 110,000 premature deaths every year. While it is difficult in these statistics to distinguish between in-door and out-door pollution effects of biomass or coal origin, studies suggest that exposed children have 2-4 times more risk of serious acute respiratory infection than unexposed children. According to the United Nations Environmental Programme, indoor air pollution caused by the widespread use of

120. Pandey (2000).

biomass as a cooking fuel is also a major contributor to the high incidence of respiratory diseases in Africa. These facts alone justify great caution in promoting policies that encourage developing countries to use more traditional biomass fuels.

This caution is further justified by considering the impact on greenhouse gas emissions. Biomass energy is generally considered to be CO_2-neutral, as long as it is consumed in a sustainable manner, i.e., the stock of biomass does not diminish. This is not the case in many developing countries, where over-consumption of biomass fuels leads to deforestation and hence to the reduction of forest-based CO_2 sinks[121]. Another aspect of biomass is that the technologies used in the developing world for converting it into useful energy have very low efficiencies, around 12-15 per cent, resulting in significant waste.

Given the usually low combustion efficiency in traditional uses of biomass, it is not necessarily true that switching from biomass to fossil fuels contributes to global warming. Studies in China and India have shown that approximately one-half of all biomass stoves, as well as most coal stoves, had substantially greater emissions of products of incomplete combustion than stoves using liquid and gaseous fuels, including 10 to 25 per cent of the embodied carbon. Measurements of the global warming potential of fuel-stoves have shown that the worst solid biomass fuel-stoves have global warming potentials ten times higher than LPG or kerosene stoves[122].

The same studies suggest that biogas stoves have global warming potentials ten times lower than stoves using LPG or kerosene[123]. This means that one of the most efficient ways to reduce the health and climate effects of biomass use in developing countries is to switch from traditional cooking stoves to modern forms of energy use.

121. This is not the case, if agricultural residues are used especially crop straws and stalks, which annually amount to 600 million tonnes in China and 320 million tonnes in India (see Sun and Gu, 2000).

122. Smith (2000). Comparison studies performed in China suggest that total greenhouse gas emissions from fuelwood stoves are only half of those from coal stoves.

123. Luo and Hulscher (2000).

Modern and more sustainable uses of biomass include biogas (see below) and biomass-fired co-generation, which can be used to provide heat and power at almost any scale, from villages to power grids. These modern technologies can use a large number of biomass fuels ranging from animal dung, crop residues, bagasse and fuelwood. Many options exist for improving the efficiency of biomass energy consumption, including improved stoves, which have been designed and distributed with varying degrees of success.

Access to and Affordability of Commercial Energy Services

Nearly 2 billion people – up to 30 per cent of the world's population, all in developing countries – have no access to electricity. Although about 300 million people have been connected to electricity sources since 1993, in the absence of adequate measures the number of people with no access to commercial energy will remain stable or continue to grow as demographic growth outpaces electrification. Commercial energy services would allow better quality of life for the poor, better education opportunities, better health conditions, better information, local industrial development and higher productivity. Access to electricity is also an indispensable first step toward the dissemination of telecommunications and digital technologies in rural areas.

Grid Extension

Access and affordability have been enhanced by government spending in this area: access through the establishment of kerosene and LPG distribution networks and through the expansion of electricity grids to rural areas, and affordability through public subsidies that have considerably lowered the prices of those commercial forms of energy.

Those public programmes, which have become increasingly difficult to administer, proved successful in some countries, but failed in others. Where programmes failed, it was often because the

Box 13

Biogas – a Special Issue

In biomass gasification, solid biomass fuels are broken down by the use of heat in an oxygen-starved environment. In China over 70 biomass gasification systems were built in the 1990s that on average deliver between 200 and 400 m^3 of gas per hour with a heat value of 5.2 MJ/m^3. These systems include a fan and a gasholder connected to a network that distributes the gas to every household for cooking. One family consumes about six cubic metres of gas per day produced from about three kilograms of straw. Each system can cover the basic energy needs of 800 to 1 600 families[124].

India has focused on biomass gasification systems for power production. A total capacity of 31 MW has been installed under a programme launched by the Ministry of Non-conventional Energy Sources. Estimates indicate that about 16,000 MW of distributed power could be generated from biomass residues. Bagasse-based co-generation is of particular significance, with an estimated possible capacity of 3,500 MW from 430 sugar mills. Projects for 222 MW have been commissioned and another 280 MW are under consideration[125].

Biogas can also be produced at the household level, with benefits for sustainability. A plant producing 2.5 m^3 biogas per day saves three tonnes of firewood and 40 litres of kerosene per year. It reduces CO_2 equivalent emissions by 4.6 tonnes per year, reduces the workload for women by three hours per day, improves health through air quality improvement and, finally, improves soil fertility and rural employment[126]. Of such small biogas plants more than 5 million are currently operating in China and more than 2.7 million in India.

The most important barrier to biomass use is variation in the quality of resources such as type of fuel, size and humidity. This means that operators must be highly skilled to ensure that the plant operates at a high load all year round. Prospects for biogas are especially good in the biomass-rich regions of South-East Asia, where wood residues, rice husks, bagasse and palm oil waste are abundant.

124. Sun and Gu (1998).
125. Gupta (1999).
126. De Castro (2000).

subsidies did not reach their targeted groups, for instance, because the poor preferred to sell the subsidy-coupons they were given rather than use them for kerosene or LPG. Rural electrification programmes, which are costly due to low population densities, combined with very low electricity prices often ruined public electricity companies and halted capacity expansion. Recognising these failures, many governments as well as multilateral funding agencies such as the World Bank have switched to promoting privatisation of national power utilities and have looked for alternative ways of providing commercial energy to the population.

Lack of access is not only an issue for rural areas. Urban residents lacking access to electricity in the fast-growing shantytowns in many mega-cities of the developing world are estimated to number some 400 million. Providing electricity to fast-growing urban populations will become ever more challenging, as the rural exodus in many developing countries is expected to continue and the world's urban population will continue to increase. Affordability of electricity is often a key issue, but there have been successful schemes in cities such as Sao Paulo and South African townships where public utilities extend their supply network to the shantytowns and practise progressive tariffs that lead to the rich cross-subsidising the poor[127].

Whether in urban or rural areas, the greatest obstacle to electrification is that populations with no access to commercial energy services are the world's poorest people. Most of them have a daily income of less than one dollar. Investment in such low-income markets requires new and innovative financing schemes, such as private-public partnerships, local co-operatives or micro-credit schemes. It is worth noting that this issue of access and affordability for commercial energy services by two billion people has attracted increasingly high-level attention worldwide. A number of important initiatives have been launched by multilateral

127. A progressive tariff is designed to subsidise the poorest groups, who consume a very small amount of electricity. Under these schemes, people with higher incomes who consume more electricity have to pay higher electricity rates.

organisations, bilateral agencies of IEA/OECD countries and non-governmental organisations.

The World Bank has been very active in this area for many years and has several programmes to address energy and poverty alleviation. The Global Environment Facility has also been actively promoting renewable energy development in the developing countries. In July 2000, leaders of the G8 countries called for international assistance to help develop markets for renewable energy in developing countries. A task force formed to identify barriers and solutions to increasing renewable energy is to report its recommendations in June 2001 (see also Chapter 6).

Off-grid Solutions

The high cost of electricity grid extension to rural areas provides opportunities for off-grid solutions. While the discussion below focuses mostly on India and China, the lessons are also applicable to other developing countries.

Small-scale hydro: Under China's Small-Scale Hydropower Programme about 20 GW of capacity have been installed during the past decades. The Programme relied mainly on domestic technology and local manpower for dam construction. The construction of the hydropower stations is funded by locally collected contributions from future beneficiaries, as well as by low interest credits provided from local and central governments. Thanks to 60,000 micro-hydro units installed in regions with no electricity grid, millions of households in rural and mountainous areas now have access to electricity. These units produce 74 TWh annually, more than one-third of the total hydropower output in China[128]. The State Development Planning Commission foresees an 8 GW increase in installed capacity of small hydropower by 2010.

India's programme for small-scale hydro is less extensive, partly due to the lack of adequate hydro resources, an estimated total

128. Total hydro capacity, including small-scale and large hydro, produced 208 TWh in 1998 or 17 per cent of total power production.

potential of 10 GW, compared to 70 GW in China. Nevertheless, installed small hydropower capacity increased threefold in the last ten years to 271 small plants with a total capacity of 217 MW and 130 additional projects under construction.

Wind power: India is well ahead of China in terms of wind power capacity. With an installed capacity over 1,000 MW, India ranks fifth in the world after Germany, USA, Denmark and Spain. China today has only about 500 MW of installed wind power capacity. In Inner Mongolia, one third of the herdsmen use about 140,000 small wind electric generators to power televisions, radios and lights[129]. In China's coastal islands wind turbines provide electricity to complement diesel groups. Both India and China also have new programmes to build large-scale wind power farms for integration into the power grid.

Solar power: Two different techniques exist to produce power from solar energy: solar photovoltaics (SolarPV) and solar thermal power (STP). Niche markets for SolarPV include stand-alone applications and small grids at remote locations. Solar thermal power is much closer to competitiveness for large on-grid applications, as long as conditions are favourable as is frequently the case in semi-arid land areas.

- *SolarPV* : In China, current SolarPV capacity is 5 MW off-grid and the government plans an additional 30 MW by 2015. In India, 15 SolarPV projects have been commissioned and ten projects of 500 kW total capacity are under installation. India is currently the third market in the world for SolarPV. As costs of SolarPV continue their downward trends, markets will expand, although it is unlikely that full competitiveness will be achieved in the next decade.
- *Solar Thermal Power*: No large-scale solar thermal power plants have been installed in any developing country so far. However, several demonstration projects are underway in India, Mexico and Morocco with funding from the Global Environment Facility

129. Lew (2000).

(GEF). All will be based on the same trough technology that is integrated into hybrid plants[130]. The GEF's solar energy programme aims at helping solar thermal power to become competitive with fossil fuels in 2005 at a price per kWh below US$ 0.09.

Solar heating: Solar heating is a more practical option for developing countries. New solar heating technology simply provides new forms of solar energy collection. Tens of thousands of solar water systems have been deployed, with a total of 3,000,000 m^2 in China and 500,000 m^2 in India. Construction of solar heated houses is developing rapidly in China and is expected to reach 350,000 hectares of roof surface this year. In India, certified solar collector companies can apply for subsidised loans and tax incentives.

Geothermal energy: Where available, geothermal energy can be used to provide heat and generate electricity for rural communities. Many developing countries, such as China and the Philippines, have significant geothermal resources. By 1995, China had more than 1 100 geothermal heating stations, providing a total of thermal capacity of 2,410 MW, the second in the world after Japan. China also had 28.6 MW of installed geothermal power generation capacity, mostly located in the Tibet area. In the Philippines, geothermal power stations accounted for 12 per cent of the total installed generating capacity in 1995, contributing 18 per cent of the country's total electricity output.

Financing Energy Projects

Financing of investment in energy projects is a particularly important issue for sustainable development in non-Member countries for at least two reasons. First, it is the necessary to build energy infrastructure to meet the large incremental energy demand in these countries. Secondly, the way those energy projects are

130. The most significant installations of such trough technology are in California, where there is 354 MW of solar parabolic trough capacity, backed up by gas-fired boilers.

built, especially the types of technology that are chosen, will have significant impact on global sustainable development. All energy infrastructure projects such as power plants, transmission and distribution networks, oil refineries or highways have a long life, lasting several decades. Each current project that does not use the cleanest most efficient technology available represents a lost opportunity for sustainable development.

According to the IEA's *World Energy Outlook 2000*, most of the energy financing needs between 1997 and 2020 will arise in the developing world, which will account for two-thirds of the total increase in world energy demand. In the electricity sector, more than half of the projected new generation capacity up to 2020 will be installed in developing countries. Of the 1,564 GW of new capacity needed in the developing world, two-thirds will be built in developing Asia. Developing countries will need to invest US$ 1.7 trillion (US$ 75 billion per year) in new generation plants alone. Similar amounts will be needed to build transmission and distribution networks. China alone will need to invest US$ 26 billion per year in new power stations over the next 20 years. Beside the enormous investment needed for the electricity sector, large sums will also be required to build associated infrastructure such as pipelines and port facilities.

In the oil sector, the *World Energy Outlook 2000* projects a growing need for capital investment in non-Member regions to expand oil production capacity. The Middle East, already the biggest exporting region, will see oil exports rise from 17 mb/d in 1997 to 41 mb/d in 2020. Timely and sustained capital investment in that region is critical to global oil security. Investment in other non-Member regions, such as Latin America, Africa and Russia is are also very important, as exports from those regions will increase significantly.

The *World Energy Outlook 2000* also projects an important need for investment in the natural gas sector to meet fast growing demand, especially in gas transportation projects to bring gas from distant reserves to consuming markets. Most gas reserves are located in transition economies, the Middle East and Africa. Globally, several

Table 12

Investment in New Power Generating Capacity By Region (US$ billion)

Region	Total Investment 1997-2020	Annual Investment
OECD	894	39
Transition Economies	319	14
Developing Countries	1,709	74
China	589	26
East Asia	290	13
Latin America	363	16
South Asia	257	11
Middle-East	114	5
Africa	96	4
World	2,922	127

Source: IEA (2000), World Energy Outlook 2000, *p. 106.*

hundreds of billions of dollars will be needed in non-Member countries each year, to build new energy supply infrastructure or to modernise existing facilities.

Building Power Infrastructures in Developing Countries

Traditionally, most developing countries relied on publicly owned companies to provide energy, and particularly electricity. This has worked in some cases, such as in Thailand during the 70s and 80s, where public funds allowed a rapid expansion of the electrification programme. However, it failed in many other countries, for example in India and many African countries, due to the low efficiency of the public utilities or the serious price distortions that deprived those utilities of opportunities for expansion.

The general trend changed in the late 80s and early 90s as inadequate energy infrastructure became a serious bottleneck for economic development. Due to strong economic growth, electricity demand out-stripped supply capacity, and the requirements to expand electricity supply largely surpassed the financial ability of existing utilities. At the same time, the ability of multilateral and bilateral lending agencies to provide funds was decreasing.

Consequently, activity in the private sector increased during the 1990s. Many governments took measures to restructure their electricity sectors to make it easier for them to access private capital. These measures included:

- Independent Power Producers (IPPs) schemes, using such models as build-operate and transfer (BOT) or build-operate and own (BOO).
- Privatisation (partial or whole) of state electricity companies, or sale of certain public assets.
- Issuance of bonds by state utilities under commercial terms and conversion of debt into equity.
- Development of domestic capital markets and listing of public utilities in those markets; and
- Purchasing of support functions such as operations and maintenance from private suppliers.

Developing countries will have to continue to rely on the private sector to meet the major part of their financial requirements for the building of energy supply infrastructure. Experience gained over more than a decade of private involvement will be valuable so that governments can define more transparent and stable regulatory frameworks, which are key for attracting investors in the global competition for foreign direct investment.

Expanding Production Capacity in Oil and Gas Producing Countries

For more than a decade, petroleum-exporting countries have periodically adjusted investment terms for outside firms who explore and produce hydrocarbons, particularly in high-cost or technically challenging areas, such as deepwater offshore West

Africa. As a result, new countries such as Equatorial Guinea and Sudan have started producing oil. Production-sharing agreements (PSA) have emerged as the most favoured type of agreement. In 1999, progress was made in Brazil, where sizeable acreage and reserves formerly held by state-owned Petrobras were licensed to private companies. Other efforts worth noting are the determination of the Nigerian government to honour the financial commitments of the state company NNPC in joint ventures with foreign companies and the apparent willingness of the Russian government to make broader use of PSA's.

Even major low-cost oil producing countries are now turning to foreign companies. Saudi Arabia's recent decision to allow foreign investment in its energy sector, except in its coveted upstream oil sector, demonstrates the awareness in oil exporting countries that they have to rely increasingly on foreign capital and know-how to sustain their production capacity. Kuwait is also considering opening its oil sector, but its parliament has questioned the transparency of the procedure proposed by the government. Libya has recently made interesting oil investment offerings. Facing competition for capital from fellow low-cost OPEC producers, Iran is also starting to realise that its "buy-back" project terms need to be sweetened.

Projected petroleum industry outlays are enormous: Nigeria reckons that US$ 35 billion are needed for its oil industry in the coming five years. Investment in Iran's oil sector will amount to US$ 10 billion in the next five years. Sustaining oil production in Russia depends heavily on attracting minimum investment of US$ 5 to 7 billion per year. In gas, up to 85 per cent of Gazprom's productive fields are in decline and need investment estimated at about US$ 2 billion per year for five years to maintain the production necessary for domestic and export markets.

However, as described in the IEA's *World Energy Outlook 2000*, it is doubtful whether such sums can be raised. Barriers to investment include poor corporate governance, erratic changes of the fiscal regime, lack of clarity of the legal base for investment projects such as disparities between regional and federal laws, continued price subsidies for residential electricity and heating as well as the non-payment of energy bills. Poor law enforcement further contributes to investor uncertainty.

Again, stable and transparent policy frameworks are crucial for attracting the inflow of private capital to finance energy projects. Additional concerns are that some of the poorest developing countries, typically in Africa, are handicapped by the small size of their markets and by their domestic consumers' poverty.

The enormous energy needs in developing countries and their future contribution to global greenhouse gas emissions will significantly impact global sustainable development. It is thus crucial that best practices and best available technologies are deployed each time a new investment is made in all elements of the energy chain – production, conversion, transportation, distribution and end-use. Investments in renewable energy projects such as wind and hydropower and in reducing the environmental impacts of large hydropower projects are essential, as are investments in the efficiency of fossil fuels technologies and advanced technologies, such as fuel cell power plants. Finally, the demand and end-use side should not be neglected. Efficient lighting, improvements in commercial and residential buildings as well as high-efficiency motors all offer good potential for energy savings.

Energy Market and Pricing Reform

Market Reform and the Social Dimension of Sustainable Development

Energy market and pricing reform pose some of the thorniest issues for sustainable development in non-Member countries, especially in the social and environmental dimensions. It is necessary first to recognise that non-Member countries have different needs in this area. In IEA/OECD countries, energy market reform concerns mostly the electricity and gas sectors. The primary objective is to introduce competition in order to improve quality of service, reduce prices to final consumers and thus contribute to economic growth. The story is very different in non-Member countries, where market reform concerns all major energy sectors: coal, oil, gas and electricity.

In transition economies, energy market reform is part of the overall economic reform programme that is changing the whole economic system from central planning to market-based economies. Even if the goal is to introduce competition and improve the quality of service, the result for final consumers is often the contrary. Market reform often leads to an increase in energy prices as a result of subsidy removal. The reform of state monopolies also frequently leads to layoffs, creating further social problems.

Market reform in the developing world is different from that in both IEA/OECD and transition economies. It is often the pre-condition for attracting any private and foreign investment to finance the expansion of energy supply capacity. As the reform process deepens, it inevitably requires the restructuring of the energy sector which often leads to layoffs and the removal of energy subsidies. For example, the China National Petroleum Corporation (CNPC) and the China Petroleum and Chemicals Corporation (SINOPEC) together have more than two million employees. The reform of those two companies could cause the layoff of one million people.

Social problems related to the loss of employment and energy price increases are significant in many developing countries. Breaking a national monopoly is difficult and is frequently slowed by industrial action or institutional obstructions. For these reasons, many countries find it difficult to create and regulate true competition among energy players.

For all the benefits that market reform brings to the energy sectors of non-Member countries, it still suffers from flaws. Often competition does not function adequately because high entry barriers prevent a sufficiently large number of new entrants to compete fully with former monopolies. Political instability, vested interests, corruption, inconsistent legal frameworks and unreliable law enforcement, administrative hurdles to investment and market pricing, distortions due to subsidies, and entrenched monopolies can all pose barriers to the efficient operation of energy markets. Finally, the lack of physical infrastructures can complicate the delivery of energy services. All these issues need to be carefully considered in order to make energy market reform a success.

Market Reform and the Environmental Dimension of Sustainable Development

Energy market and pricing reform in non-Member countries also has a profound impact on the environmental dimension of sustainable development. Here again, a clear distinction can be made. In IEA/OECD countries it is not clear if market reform and deregulation can produce positive effects on the environment. In transition economies and developing countries it is much clearer that market and pricing reform is necessary to achieve better environmental performance in the energy sector.

In several developing countries, market reform has had a positive environmental impact in the following ways:

- Compared to the old state-owned utilities, which often failed to respect environmental regulations, private and foreign actors provide better technologies that reduce local and global environmental emissions. Conscious of the future tightening of environmental standards, they often provide technologies that exceed current environmental norms.
- Private actors can bring better environmental management skills and know-how along with their investments.
- Some governments have used the opportunity of market reform to improve outdated environmental regulations, introducing a number of economic instruments.
- Privatised companies are much more responsive to environmental regulations than public monopolies; and

- Environmental enforcement is easier *vis-à-vis* a private company than a public utility.

However, energy market reform is not a one-way bet. Price competition may be detrimental to long-term investment or may impede the development of costlier renewable energies having high front-end costs. Traditional demand-side management (DSM) programmes may also be less attractive to utilities when they are privatised.

Despite the initial benefits market reform can bring in the environmental dimension, some long-term issues require attention. Over the long-term, market forces alone will not be able to cope adequately with issues such as environmental and health externalities. It is thus important to establish adequate frameworks to balance the expectations of investors for a return on their investments and the longer-term goals of environmental sustainability. Other than assuring the rule of law, this includes incentives to develop non-electrified regions, formulating and enforcing environmental regulations as well as investing in renewable energy sources.

The Issue of Energy Subsidies

As noted in Chapter 2, energy price reform and energy subsidy removal are among the thorniest reform issues in non-Member countries due to the hardships from lifting subsidies and hiking prices. Reducing or scrapping energy subsidies is often deeply unpopular and can sometimes jeopardise political stability.

Energy subsidies in non-Member countries are far larger than in IEA/OECD countries. Estimates of worldwide energy subsidies vary from US$ 100 to 150 billion, roughly three-quarter of which are spent by non-Member countries. Whereas IEA/OECD countries tend to subsidise producers, developing countries tend to subsidise consumers. Transition economies subsidise both consumers and producers.

Energy subsidies have created serious distortions in energy pricing in non-Member countries. They also have led to the poor financial performance of many state-owned energy companies in developing and transition economies. This poor performance in turn has reduced the ability of those companies to invest in new generating capacity or grid extension, especially to provide service to consumers without access to commercial energy.

Price-distortions have accelerated the depletion of domestic resources and distorted industrial and infrastructure development. For example, in India the heavy subsidy provided for diesel oil has led to a "dieselisation" of the Indian economy[131]. Its consequences include: heavy air pollution in major cities, very low end-use efficiency, misdirected infrastructure development, wasteful use of government money and a growing threat to India's energy security and macro-economic stability.

Price reform and the removal of subsidies are also critical for transition economies. In Russia, artificially low domestic energy prices weaken incentives for adopting new energy efficient technologies and deprive the energy sector of much-needed resources for investment. Payment of energy bills, however, is improving slowly after the government narrowed the list of customer categories protected from disconnection in 1997.

The IEA carried out a study on the energy subsidies of eight large energy-consuming non-Member countries in 1999[132]. It revealed that pervasive energy subsidies exist in those countries, from 6.4 per cent of the cost of production in South Africa to 80.4 per cent in Iran. It concluded that the abolition of those pervasive subsidies in those countries would:

- reduce primary energy consumption by 13 per cent;
- increase GDP through higher economic efficiency by almost 1 per cent;

131. Eighty per cent of road vehicles in India run on diesel. Diesel-generators for industrial and commercial establishments and wealthy urban households have proliferated. For freight road transport is used instead of transport by rail. For more details, see Chen (1997), "The Dieselisation of the Indian Economy".

132. China, India, Indonesia, Iran, Kazakhstan, Russia, South Africa and Venezuela. See IEA (1999), Looking at Energy Subsidies: Getting the Prices Right.

- lower CO_2 emissions by 16 per cent; and
- produce domestic environmental benefits, including reduced local air pollution.

Globally, those benefits would reduce worldwide energy consumption by 3.5 per cent and CO_2 emissions by 4.6 per cent.

More and more governments have come to understand the counterproductive effects of energy subsidies. Pressed by budgetary constraints or encouraged by outside forces, many countries have cautiously brought energy prices closer to cost recovery, but several have faced strikes or social strife as a result (for example, Indonesia, Nigeria, Egypt, Venezuela, India or Ecuador). Much progress has been made in China, where domestic oil prices are now fully linked to world prices and electricity prices are close to international levels. Coal subsidies have been much reduced, and more than 30,000 small coalmines were closed. Several countries have also adopted schemes to buffer the social hardships of subsidy removal, typically through targeted subsidy of minimum electricity supplies for the poor.

Technology Transfer in the Context of Sustainable Development

Improved energy technologies contribute to sustainable development in non-Member countries in two major ways. First, they allow cleaner energy production. Second, they improve the quality of life through expanded energy services like transport, heating and lighting. Energy services such as refrigeration and sterilisation also contribute to improvements in public health and reduce infant mortality. However, there is no guarantee that new technologies will appear when they are needed or at a price that reflects all the benefits associated with them. Government action to correct these market failures is therefore often appropriate.

International co-operation can assist non-Member countries in accelerating the adoption of new and more efficient energy technologies. Two major fields for co-operation with non-Member countries are information about technologies that allow better trade-offs between expanded energy services and cleaner energy use, and follow-up programmes to assist in the application of these technologies. It is particularly important for technology co-operation activities to be integrated with the full range of development co-operation activities, especially energy-related market and policy reforms, institutional restructuring and project financing, but also with core activities, such as poverty alleviation.

Measures to Accelerate Technology Transfer

In assisting developing countries and promoting technology transfer *top-down activities* are distinct from *bottom-up activities*. Top-down activities centre upon identifying and setting appropriate national framework conditions. For the most part, the appropriate framework conditions for encouraging application of cleaner technologies are the same as those that encourage the development of an efficient, effective energy sector. Actions by OECD countries to encourage the application of cleaner energy technologies cannot take place in a vacuum. Proposals need to be assessed against non-Member countries' own strategies for sustainable development. But few non-Member countries have cohesive policies in this area. Here is where IEA and OECD can contribute.

Bilateral and multilateral development assistance agencies such as the World Bank, UNDP, export credit agencies of OECD countries and others have been providing assistance in the formulation and implementation of policy frameworks. Examples include: identifying priority needs for sectoral reform, designing market incentives and creating the necessary institutional frameworks. All too often, however, such attempts have been concerned with developing sectoral policy frameworks without reference to sustainable development.

The main actor in technology innovation, diffusion and application is the private sector. It should be involved at an early stage in policy formulation and in the design of regulations and enforcement mechanisms. Efficient channels of communication and greater collaboration between industry and government are important instruments in this regard and in an increasing number of countries, business and industry associations are improving their co-operation with government. Finally, enhancing public awareness of the health and other impacts of pollution and the need for a transition toward sustainable development are important factors in promoting effective policy frameworks.

The bottom-up approach, on the other hand, focuses on individual projects. It involves working with communities, local firms and multinational investors. A typical energy-related project using official development assistance is the Asian Development Bank's project to increase the efficiency of the district heating system in Ulan Bator. In China, an ADB project provides loan financing for the construction of three wind farms with a total power generation capacity of 78 MW. This project also receives technical assistance from the Global Environment Facility (GEF) to promote further development of wind-based power generation in China.

Loan financing and technical assistance are the two main benefits that external agencies bring to projects in developing countries. Technical assistance is generally required to set the scope for potential projects and fine-tune institutional arrangements. Without the groundwork of technical assistance much loan finance, including private sector finance, would never be realised.

Capacity Development

The principal constraints to the rapid diffusion of cleaner technologies in developing countries relate to a lack of institutional and managerial capacities. Support for the dissemination of technological know-how must concentrate on developing the necessary human, scientific, organisational and institutional capabilities to underpin the long-term application of new technologies. Capacity development is a long-term process not a finite product. Effective technology co-operation may require commitments for support that go beyond the normal planning horizon of three to five years. Policy and institutional sector reform take even longer, in some cases 10-20 years.

Capacity development for cleaner technologies encompasses the whole range of assistance activities designed to develop the skills, knowledge and technical know-how to allow developing countries to adopt cleaner production technologies and adapt them to their needs. Some of the main instruments of technical co-operation include: training for engineers and other groups of personnel in the in the private sector; demonstration and pilot projects; and contributions to science, research and technological development.

What More Can Be Done?

Work related to technology deployment in non-Member countries should be consistent with broader work with non-Member countries on sustainable development. All else being equal, priority should be given to non-Member countries with the highest energy use. However, for certain classes of technology, markets might evolve more quickly in other countries, before achieving levels of performance where global market penetration occurs.

Whether undertaken by the IEA, the World Bank or other international organisations, such work needs to adhere to a number of informal criteria in order to deliver tangible benefits for the transfer of technologies that can contribute to the three dimensions of sustainable development. International organisations should avoiding duplication and overlap. Technology co-operation efforts of Member countries and multilateral donor organisations in key non-Member countries should be pooled.

A particular important issue in this context is the engagement of industry through well-defined activities specified by the participants themselves. This requires a respect for commercial confidentiality that is best insured by voluntary "opt-in" for different activities. At the same time, a certain degree of government involvement is necessary since the purely commercial activities of private firms and consultants would not maximise the benefits for sustainable

Box 14

IEA Initiatives in Technology Transfer

The IEA – together with a number of other international organisations — is a source of information on cleaner energy technologies for non-Member countries that choose to participate in the IEA Implementing Agreements. *Currently, 11 non-Member countries participate in 26 of the 40 active Implementing Agreements. In total, non-Member countries account for 42 participants, compared with a total of 418 participants from IEA Member countries. The most active non-Member countries are the Republic of Korea (participating in 11 Implementing Agreements), Mexico (seven) and Russia (seven). The participation of non-Member countries in Implementing Agreements has more than doubled over the past three years. These Agreements are based on the principle of equitable sharing of costs and benefits and the accumulated information is not generally available to non-Participants. Implementing Agreements can assist in exploring opportunities for third-party financing of participation of non-Member countries.*

The Climate Technology Initiative (CTI) – an initiative of IEA/OECD Member countries, whose Secretariat is hosted by the IEA – aims more specifically to increase the effectiveness of environmentally sound technology and practices through activities in three primary areas: capacity building, technical assistance and mechanisms for technology transfer. Its mission is to promote international co-operation for accelerated development and diffusion of climate-friendly technology and practices in line with the objectives of the UN Framework Convention on Climate Change (UNFCCC). CTI works with several partner organisations, including the World Business Council for Sustainable Development (WBCSD), the International Standards Organization, E-7 Network of Expertise for the Global Environment, Versailles Agreement on Advanced Materials and Standards (VAMAS), the Edison Electric Institute, IEA/GREENTIE, and the International Co-generation Alliance (ICA). During the year 2000 alone CTI sponsored seminars and information diffusion activities in Asia, Southern Africa, Eastern Europe and Latin America.

development. Technology transfer requires co-ordinated efforts from a multitude of actors: governments of developing countries, governments of IEA/OECD countries, the private sector and international financial institutions (IFIs)[133].

Such international co-ordination would advance a framework for technology collaboration. The framework could include co-ordination on national needs assessments, prioritisation exercises, training, institutional strengthening and financing. As many stake-holders as possible from both donor and developing countries should be involved, including economic, energy and environment ministries, development agencies, local governments and private investors.

At all times, the deployment of promising technologies should take account of the concrete requirements of non-Member countries. These requirements need to be established in an open and equitable dialogue. Building such partnerships is greatly helped by making a medium to long-term commitment to work with particular non-Member countries, which will then be encouraged to bring their own resources to the dialogue. There is a need to move beyond limited provision of information and scattered, one-off workshops and other events. Best results for sustainable development are achieved by assisting non-Member countries through longer-term commitments to co-operate in all three areas of technology transfer: building framework conditions, individual projects and the development of local capacities (see also Box 15 below).

Striking a Balance Between Economic Development and Environmental Protection

A final issue for energy and sustainable development in non-Member countries is the need to strike a balance between economic development and environmental protection. This is at the heart of sustainable development. For decades, many developing countries have focused excessively on economic growth

133. *See, for instance, IEA (2000g),* Energy Technology and Climate Change – A Call to Action.

Box 15

Ten Concrete Suggestions for Furthering Technology Transfer

1. *Technology co-operation should be consistent with the development priorities of developing countries.*
2. *Technology co-operation should be an integral part of development co-operation, from technical assistance on energy sector institutional reform, to poverty alleviation programmes.*
3. *Collaborative activities would help to avoid duplication and overlap.*
4. *Activities should seek to foster institution building and to increase capacity in non-Member countries.*
5. *Participatory approaches should be adopted to ensure that activities are demand-driven.*
6. *Individual activities should build on national strategies for sustainable development.*
7. *One-off, isolated events are not appropriate. Engagement with individual non-Member countries should be through medium- to long-term programmes.*
8. *Most developing countries, and all large ones, are on unsustainable development paths. International organisations such as the IEA need to play a role in order to avoid economic and environmental risks.*
9. *The key measure of success should be the creation of sustainable markets for cleaner energy technologies. Partnerships with the private sector are vital in this context*
10. *Because of their leading role, private businesses should be included in collaborative processes at an early stage.*

as their first priority and more or less ignored the need to protect the local and global environment. But as their economies grow and environmental degradation increasingly affects the lives of their people, they become increasingly aware of the need for environmental protection, first local, then global.

As a result of decades of industrialisation without attention to the environment, the top ten most air-polluted cities in the world are in developing countries according to the World Health Organisation. Among them, China and India hold a sad record. Major cities in China have frequently ranked high, mainly due to widespread coal use. More than 500 Chinese cities are reported to have air quality below the WHO criteria. Particulate and sulphur levels exceed WHO and Chinese standards by two to five times. Chronic obstructive pulmonary disease is the leading cause of death in China, partly because of outdoor and indoor pollution. According to a World Bank study, 200,000 people die in China each year because of excessive air pollution.

In addition to the premature death of people, air pollution has caused severe damage to the Chinese economy. According to Chinese experts, environmental pollution was directly responsible for economic damage at the rate of about 7 per cent of China's GDP in the early 1990s. If the degradation of environmental quality as well as expenditures on pollution control were included, the total cost to the society would amount to 10-15 per cent of the country's GDP. Air pollution was responsible for about one-third of the total cost.

Another serious problem in China is acid rain due to SO_2 emissions. Caused mainly by coal combustion, acid rain now affects 30 per cent of China's agricultural land and has begun to reach Korea and Japan. In 1995, the direct economic loss due to acid rain alone in China amounted to US$ 14 billion, equivalent to 2 per cent of the country's GDP. Total economic cost due to air pollution in 1995 amounted to US$ 48 billion[134].

India also has many of the world's most air-polluted cities, such as Delhi and Calcutta. The air quality data for India's four largest

134. ERI (2000).

urban conglomerates indicate that ambient levels of suspended particulate matter (SPM) considerably exceed WHO standards as well as national air quality standards set by the Indian Central Pollution Control Board. A study on benzene concentrations in Delhi conducted by the Netherlands Institute for Applied Research showed that the air-born benzene density on an open road in Delhi was six times higher than in a traffic tunnel in Rotterdam. Perhaps more telling is the comparison with a similarly large city like Cairo: benzene concentrations in Delhi are about three times as high than in Cairo. These levels of carcinogenic benzene are related to the large number of old two-stroke engines and the poor quality of diesel and lubricating oil.

Delhi houses 13 million people. Every day, 2,000 tonnes of air pollutants are emitted into the atmosphere. Local pollution is responsible for 7,500 deaths per year. The situation is similar in many mega-cities of Southeast Asia such as Bangkok and Jakarta. Quantitative estimates showed that the environmental cost of air and water pollution in Jakarta and Bangkok exceed US$ 1 and 2 billion per year respectively, around 8 per cent of the total income of these cities.

Increasingly, governments of non-Member countries have realised that this pattern of growth is unsustainable, not only for future generations, but also for today's. The costs in terms of public health expenditures, increased mortality, losses in resource-based sectors as well as in terms of the irreversible loss of biodiversity and the decline in overall environmental quality are too large to be ignored any longer.

Responding to concerns that environmental degradation destroys value in the order of 10-15 per cent of the GDP, the Chinese government took several measures to improve environmental protection. The authorities closed down over 30,000 small coal mines over the last five years. For years the central government has also been trying to prohibit the construction of small coal-fired power plants and has advocated that existing ones be taken out of service. Coal-fired power plants of 50 MW and under tend to be less efficient and more highly polluting per kW of generating capacity than larger plants. With growth in demand slowing in recent years and sensitivity to environmental pollution gaining strength, the government has renewed its commitment to closing plants. According to one report, 2,840 MW of plants smaller than 100 MW were closed in 1997 and 1998 and a further 1,800 MW were slated for closure in 1999. An additional 7,740 MW were to be eliminated in 2000[135].

All across China, and particularly in the wealthy coastal provinces, cities and towns are gradually becoming stricter about enforcing limits to pollutant emissions. When local administrations are supportive, environmental protection agencies can levy significant emissions fees and fines or mandate process changes. In some cases this has resulted in the installation of pollution control equipment. In other cases, urban factories were forced to either move to rural locations or be shut down altogether. Many cities also banned the use of poor quality coal within defined zones.

In a 1996 decision, the State Council had ordered all industrial enterprises in China to meet their environmental emissions standards by the end of 2000 or face closure. By the end of July 2000, according to the State Environmental Protection Administration, 90 per cent of the country's 238,000 industrial enterprises, but only two-thirds of the 620 largest state-owned ones, had met the standards. In August 2000, the authorities of the northern city of Shenyang shut down for the first time a large enterprise (Shenyang Smelter) because its polluting emissions costing 20,000 people their job. The case was held up by the Environmental Protection Administration as a threatening example: "If an enterprise doesn't eliminate pollution, pollution will eliminate the enterprise".

Similar actions were also taken in India, although to a less drastic degree. Urban authorities have taken actions to improve ambient air quality by relocating polluting industries from high population

135. Sinton and Fridley (2000).

density areas to lower population density areas. For example, this includes Delhi, which has experienced a massive growth in small-scale industries in the last 15 years and has been directed by the Supreme Court to relocate 114 highly polluting stone crushers. Consequently many of these offenders have moved into the neighbouring state of Haryana[136]. However, the decision by the Indian government in November 2000 to order the closure of 55,000 medium and small enterprises in Delhi as an anti-pollution measure caused violent strikes by the 600,000 people employed by these enterprises.

Many of the measures that are aimed at local pollution also have consequences for global emissions. Reducing emissions of local air pollutants contributes to the reduction of global greenhouse gas emissions. For example, Chinese authorities claim that the closure of coal mines and small coal-fired power plants, structural change in the economy and efficiency improvements have reduced China's CO_2 emissions by 20 per cent since 1996.

Specific Challenges and Global Implications

This chapter discussed six specific challenges facing non-Member countries in regard to sustainable development in their energy sectors. Each of these has strong implications for the global community:

The *use of biomass energy* in developing countries affects the global coverage of forests, land-use and soil quality. It is very important to promote sustainable ways of using biomass energy or even to replace it with commercial fuels, such as LPG.

Access to and affordability of commercial energy services is vital for the prosperity of several billion people. The way those people improve their life and their environment is critical to the sustainability of the whole world.

136. WWF (1995)

The *financing of and investment in energy infrastructure projects* also directly affect global strategy for sustainable development. Expanding energy supply capacity is essential to meet growing energy demand. The way energy projects are financed, especially what type of technology is chosen, will greatly affect the future pattern of environmental emissions.

Energy market and pricing reform, which is mainly driven by economic objectives, also strongly affects the social and environmental dimensions of sustainable development. International involvement, such as membership in the World Trade Organisation, can accelerate reforms. The impact of market reform in non-Member countries goes far beyond their borders. IEA/OECD countries have important experiences to share with developing countries on how to safeguard competition, environmental and social objectives in liberalised energy markets.

Technology transfer requires long-term commitments by bilateral and international organisations to create appropriate institutional infrastructures, to promote project-based co-operation and to initiate capacity building. Particular emphasis needs to be placed on the avoidance of overlap and the involvement of private sector participants.

Finally, the *balance between environment and economic development* is at the heart of global sustainable development. Developing countries need to strike a right balance between these two objectives for their own sake. However, IEA/OECD countries have an interest assisting them in this process as they stand to benefit from diminished regional pollution such as acid rain and from reductions in global greenhouse gas emissions.

While non-Member countries must be engaged in the efforts of IEA/OECD countries to achieve sustainable development, this engagement has to take into account the specific situations in non-Member countries. Only by proposing measures that also address the specific challenges faced by those countries can they be effectively drawn into global efforts. And the most pressing environmental challenge in many developing countries remains local environmental pollution.

An important way of engaging non-Member countries is to share knowledge of policy lessons, experiences and best energy practices, recognising that this knowledge needs to be adapted to the specific situations in each country. An important element of global co-operation is the so-called "South-South dialogue", whereby developing countries share experiences and lessons with each other. For example, India and China could exchange experiences and lessons on their respective national programmes for improved stoves. Despite imperfections in these programmes, the mass diffusion of reliable and truly improved stoves would have far greater positive environmental impacts than efforts to disseminate expensive and sophisticated renewable technologies.

On the other hand, developing countries also should be encouraged to "leap-frog" the industrialised world in their sustainable development strategies when the opportunity arises. Because they have the opportunity to learn from the experiences of industrialised countries, they can avoid past errors and choose more sustainable development paths from the start. With better knowledge about the economic and environmental risks represented by certain technological choices, they can make sounder investment decisions. Today a wide range of low-cost technological options can be carefully assessed to match them closely to environmental objectives. These opportunities need to be exploited to create incentives for the deployment of sustainable technologies and practices.

POLICY CONCLUSIONS

Toward a Framework for Action

A forward-looking policy, integrating the economic, environmental and social pillars of sustainable development and including appropriate risk management is the principal feature of a coherent approach to sustainable development in the energy sector. This approach has been developed on the preceding pages. The concrete policy options that follow from such an approach are summed up in the recently adopted IEA Statement on Sustainable Development (see box below).

The IEA Statement on Sustainable Development enlarges the *Shared Goals* to which IEA Energy Ministers committed themselves in 1993. The main goals that were put forward then – diversification, energy efficiency and cost-reflective pricing – remain valid today.

Box 16

IEA Statement on Sustainable Development

Energy and Sustainability: Key Features

1. Energy has deep and broad relationships with each of the three pillars of sustainable development — the economy, the environment and social welfare. It remains a strategic commodity: social and economic development can be attained only so long as a secure, reliable and affordable supply of energy is ensured. Energy services help to fulfil basic needs such as food and shelter. They contribute to social development by improving education and public health and, overall, help alleviate poverty. Access to modern energy services can be environmentally beneficial, for example, by reducing deforestation and decreasing pollution through more efficient energy use.

2. These different dimensions are intrinsically linked. Sustainable development is dependent upon balancing the interplay of policies and their effective implementation to achieve economic, environmental and social needs.

Economic growth requires a secure and reliable energy supply, but is sustainable only if it does not threaten the environment or social welfare. Environmental quality is more readily protected if basic economic needs are also met, and social development needs both economic growth and a healthy environment. Sometimes the policies are mutually reinforcing and sometimes they are in conflict, and trade-offs will often need to be made. Lower fuel prices widen access to energy, but also encourage inefficient utilisation of energy resources and accelerated resource depletion. Conversely, if energy prices are raised too quickly in an effort to combat environmental concerns, energy may become too costly and thus placed beyond the reach of those who need it most.

3. The path to a more sustainable energy future is not static. It must be continuously redefined and rebalanced with revised forecasts, reassessment of progress, identification of new problems and the development of new technical solutions and technologies. All countries – developed and developing – will need to design their own policy mix; it is clear that national circumstances will affect the scope for action and the appropriate policy choices in and between countries. The policy makers' task is to assess the risks to, and from, today's energy systems. They must determine what changes would advance economic, social and environmental objectives. Policymakers must look to the long term, taking action today to avoid longer-term social, economic or environmental disruptions, while retaining flexibility to alter course when the existing path proves to be unsustainable.

Are we on a sustainable energy path? Not unless we make considerable changes.

4. Projecting the current energy situation and energy policies into the future suggests growing pressures on the global economy and the environment. Governments need to develop policies to address the projected 57% increase in the predominantly fossil-fuel based global energy demand over the next 20 years. Governments also need to take action to modify longer-term trends in greenhouse gas emissions within the framework of the United Nations Framework Convention on Climate Change. Policies will need to take into account that the energy demand of non-OECD countries

will soon surpass that of OECD countries, and that developed countries' already high levels of energy demand will continue their upward trend. Policies will also need to address potential decline in energy security as the sources of oil and gas production become more concentrated in regions of geopolitical uncertainty. Capital markets and governments will need to seek ways to mobilise the enormous resources to meet growing energy needs.

5. Sustainability demands that we seek to change present trends. The challenge is to fuel world-wide economic growth with a secure and reliable energy supply, without despoiling our environment. It is possible. Energy supply needs to be further de-carbonised, diversified and the energy intensity of economic growth reduced. Global energy security can be enhanced through collective efforts and efficient but well-regulated markets can make energy affordable.

Towards a Solution

6. The transition to a sustainable energy future will be complex and will take time. We need to change not only the structure of the energy sector, but also behaviour in our societies and economies.

7. Consistent with the Shared Goals of the International Energy Agency which call for policies that balance energy security, economic growth and environmental protection, Member Governments of the IEA seek to create the conditions in which the energy sectors of their economies can make the fullest possible contribution to sustainable development. These include:
   - *Safeguarding energy supplies* through diversification and through co-ordination of the use of flexible response mechanisms in the event of supply disruptions.
   - *Promoting further improvements in energy efficiency*, along with further development and diffusion of non-fossil fuel technologies, including renewable energies.
   - *Ensuring that energy markets operate in a competitive and transparent manner with minimum distortions.* As prices shape behaviour and technology, price signals reflecting full costs should reach consumers. This will entail the

gradual elimination of environmentally harmful subsidies, and internalisation of externalities (such as environmental costs and benefits), ideally through the use of market-based instruments. Reduction of trade and tariff barriers will help markets operate openly and competitively and improve confidence in the marketplace.

- *Creating a stable framework for decision-making, one that includes clear signals to the market.* Incentives, regulatory measures and standards will be needed to stimulate sustainable choices in a marketplace that is still economically imperfect.
- *Continuing to liberalise energy markets with frameworks to protect the environment and enhance social welfare.* These frameworks should be stable and predictable, and promote open and competitive energy infrastructure.
- *Encouraging the systematic introduction of the best technological solutions where energy investments are made.* Capital stock turnover and new additions to the capital stock offer important opportunities for increasing the use of cleaner, more efficient technology.
- *Participating in a global effort to provide electricity for those currently without access*, through the development and diffusion of technologies and the development of stable legal, fiscal and energy policy frameworks, particularly in developing countries, that stimulate the flow of private capital.
- *Ensuring high safety standards* in the operation and maintenance of energy equipment, plants and infrastructure, and putting in place appropriate mechanisms to respond to potential accident or failure.
- *Sponsoring energy research and development, information exchange (including data and statistics) and dissemination* with a view to encouraging commercial applications and changes in consumer behaviour. Transparent decision making processes are required with broad policy-maker participation — for example, from transport, industry, trade, environment and finance — as well as wider stakeholder involvement.

One challenge in the development of sustainable energy policies is the strong – but complex – linkages to which they are subjected. These linkages can make the definition of precise sub-goals difficult. Efforts to reduce CO_2 emissions, for instance, are closely linked to the competitiveness of renewable energy and energy efficiency. All three have an effect on energy demand and supply security. Market liberalisation has both negative and positive implications for technology and environmental performance. Certain policy instruments, especially market-based instruments such as a carbon tax, have multiple effects on more than one dimension of sustainable development.

Such interactions increase uncertainty. Uncertainty is further heightened by the rapid structural and technological change that is sweeping through the energy sector. Globalisation and the increasing interconnection of OECD countries with non-Member countries, both the developing countries and economies in transition, have become an inescapable reality. At the same time, privatisation and market liberalisation increase the number of decision-makers further, creating both opportunities and challenges for sustainable development.

Many of the actors in the new markets do not automatically strive to achieve sustainability objectives. Recent experiences in liberalised electricity markets, for instance, have shown that private market participants can underestimate the importance of security and reliability of supply, including their true value to energy consumers and society at large. In short, structural change will only deliver its full potential for sustainability if wisely designed and implemented, within a framework which provides for any necessary guidance or direction to ensure attainment of public policy objectives.

Identifying Concrete Steps Toward Sustainability

This study has identified the broad outlines of policies that could promote sustainable development in the energy sector.

- Policies need to strive for balance among the economic, the social and the environmental dimension and account for the interactions between them. As part of this balance, policy integration is a necessary condition for success in the energy sector. Information sharing and consultation with responsible decision-makers in the economic, environmental and social area need to be further developed.
- Policies need to manage the different dimensions of risk that can pose threats to sustainable development of, and stemming from, the energy sector. Policies must recognise that unexpected shocks will certainly arrive and that there is no such thing as "zero-risk". Thus, it will be critical to maintain response flexibility in order to cope with such events as they arise. Managing supply security risk, technical or project-related risk, as well as climate change risk, are the three most urgent tasks in this respect.
- Policies need to contribute to the management of social risk by promoting reliable and affordable access to energy – especially in developing countries.

Specific conclusions closely track these broad outlines. Thus, for example, the issue of managing risk is highlighted in the climate change discussion. Solutions, including the use of cost-effective, efficient market mechanisms are proposed to address the problem. The broader issue of the participation of developing countries in the future energy system is also discussed. From climate change and the security of energy supplies to energy investments and technology transfer, the success for sustainable development hinges on the co-operation of industrialised and developing countries[137]. This requires the patient construction of international frameworks in which new forms of co-operation can take place that take account of the prerogatives and the sensitivities of all sides involved.

137. It is important to understand to what extent this is not a one-way relationship. The discussion of "learning curves", for instance, points out that in the area of renewable energy technologies producers in OECD countries depend on non-Member countries, as well as on their home countries, to provide the markets for production volumes that can then generate cost reductions for all involved.

One more theme also consistently emerges from the analysis: one of the most important criteria for policymaking in the energy sector is *coherence*. The consistency of broad policy orientations concerning different aspects of sustainable development as well as their continuity through time is an important condition for success. The demand for coherence and for the long-term orientation of policymaking goes hand in hand with an insistence on establishing credibility and trust between different stakeholders. Coherence, credibility and trust are also pre-conditions for improving co-operation at the national and international levels between governments, industry, civil society and international organisations. Coherence is also the basis for a transparent and convincing information policy *vis-à-vis* a public that needs to be prepared to make the difficult tradeoffs required to progress toward sustainable development.

Under these general conditions, concrete policies can be developed around four axes.

- Getting market signals right so that prices reflect the true costs of producing and consuming energy in undistorted markets.
- Making trade and investment compatible with environmental and social criteria by emphasising the reduction of environmental side-effects and the extension of access to electricity.
- Providing support for innovation and technological change when public goods are concerned, especially in the areas of energy efficiency and renewables.
- Creating broad institutions that foster co-operation, research, learning and transparency.

BIBLIOGRAPHY

AMALBERTI, René (1999), "Les effets pervers de l'ultra-sécurité", *La recherche* 319, p. 66.

AMALBERTI, René (2000), "Dans les entreprises le risque zéro n'existe pas", *Le Monde*, 15/8/2000.

ARROW, Kenneth J. (1966), "Discounting and public investment criteria" in Alan V. Kneese and S. C. Smith, Eds., *Water Research*, Johns Hopkins University Press for Resources for the Future, Baltimore.

ARROW, Kenneth J. and William R. CLINE, Karl-Göran MALER, Mohan MUNASINGHE, R. SQUITIERI and Joseph STIGLITZ (1996), "Intergenerational equity, discounting and economic efficiency" in J.p. Bruce, H. Lee E. F. Haites, Eds., *Climate Change 1995 – Economic and Social Dimensions of Climate Change*, Contribution of Working Group III to the Second Assessment Report of the Intergovernmental Panel on Climate Change, Cambridge University Press, pp. 125-144.

AWERBUCH, Shimon (2000), "Getting It Right: The Application of Portfolio Theory to Renewables Standards", *Public Utilities Fortnightly*, 15 February 2000, pp. 44-55.

BARNETT, Andrew (1999), "A Review of the Renewable Energy Activities of the UNDP/World Bank Energy Sector Management Assistance Program, 1993-1998," ESMAP Report 223/99, Washington, DC.

BARON, Richard (1999a), *An Assessment of Liability Rules for International Greenhouse Gas Emissions Trading*, IEA Information Paper, Paris.

BARON, Richard (1999b), *Market Power and Market Access in International Greenhouse Gas Emission Trading*, IEA Information Paper, Paris.

BARON, Richard (2000a): *International Emission Trading – A Real Time Simulation*, IEA Information Paper, Paris, available at: http://www.iea.org/envissu/cop6/emistr.pdf.

BARON, Richard (2000b), *Market Access Issues in International GHG Emission Trading*, IEA Information Paper, Paris.

BATTELLE MEMORIAL INSTITUTE (1998), *China's Electric Power Options: An Analysis Of Economic and Environmental Costs*, Washington DC.

BHATTARAI, T.N. (1998), *Charcoal and its Socio-Economic Importance in Asia: Prospects for Promotion*, Regional Wood Energy Development Programme, Bangkok.

BOHI, D.R. (1983), *What Causes Oil Price Shocks?* Discussion Paper D-82S, Resources for the Future, Washington, DC.

BOHI, D.R. (1989), *Energy Price Shocks and Macroeconomic Performance*, Resources for the Future, Washington, DC.

BOHI, D.R. and Michael A. TOMAN (1996), *The Economics of Energy Security*, Kluwer Academic Publishers, Boston.

BROWER, M. (1995), "Comments on Stirling's Diversity and Ignorance in Electricity Investment", *Energy Policy* 23: 115-116.

CABRAAL, A., M. COSGROVE-DAVIES, and L. SCHAEFFER (1998), "Accelerating Sustainable Photovoltaic Market Development", *Progress in Photovoltaics: Research and Applications* 6: 297-306.

CARLSSON, Bo and R. STANKIEWICZ (1991), "On the Nature, Function and Composition of Multinational Enterprise" in Richard D. Robinson, Ed., *The International Transfer of Technology: A Book of Readings*, Ballinger.

CHEN, Xavier (1997), "The Dieselisation of the Indian Economy", *Geopolitics of Energy*, pp. 9-11.

COHEN, Joel E. (1995), *How Many People Can the Earth Support?* Norton, New York.

DE CASTRO, J. (2000), "Gasification and Biogas Technologies" in FAO (2000), pp. 56-61.

DODGSON, Mark and John BESSANT (1996), *Effective Innovation Policy: A New Approach*, International Thomson Business Press.

DOE (1990), "Report of the NES Oil Externality Subgroup", Draft Manuscript, United States Department of Energy, Washington, DC.

DOE (1997), *Electricity Prices in a Competitive Environment: Marginal Cost Pricing of Generation Services and Financial Status of Electric Utilities*, Report No. DOE/EIA-0614, United States Department of Energy, Washington, DC.

DUNN, Seth and Christopher FLAVIN (2000), "Sizing up Micropower" in *State of the World 2000*, Worldwatch Institute, Washington, DC, pp. 142-161.

ED (1996), "Stranded Cost Speculation: Rudden Says $65 Billion at Risk", *The Energy Daily*, 29 July 1996.

ERI (2000), *China's Energy and Environment Problems and the Necessity for International Co-operation in North-East Asia*, Research Report, Energy Research Institute, Beijing.

EUROPEAN COMMISSION (1995), *ExternE: Externalities of Energy*, Brussels.

EUROPEAN COMMISSION (2000a), *Climate Negotiation and Emission Trading: Economic Insights from European Models*, Synthesis Report, Seminar, 29-30 August 2000, Brussels.

EUROPEAN COMMISSION (2000b), *Green Paper on Greenhouse Gas Emissions Trading within the European Union*, COM(00)87, Brussels.

FAO REGIONAL WOOD ENERGY DEVELOPMENT PROGRAMME IN ASIA (2000), *Wood Energy, Climate and Health*, Proceedings of the International Expert Consultation, 7-9 October 1999 in Phuket, Thailand, Field document n° 58, Bangkok.

FLAVIN, Christopher and Seth DUNN (1999), "A new energy paradigm for the 21st century", *Journal of International Affairs* 53(1): 167-190.

FLOWER, L. (1997), "Renewables for Sustainable Village Power", Paper presented at the International Conference of Village Electrification through Renewable Energy, New Delhi, India, 3-5 March 1997, available at http://www.rsvp.nrel.gov.

FLOWER, L. (1998), "Lessons Learned – NREL Village Power Program", National Renewable Energy Laboratory, Golden, CO, available at http://www.rsvp.nrel.gov

GAO (1996), *Evaluating US Vulnerability to Oil Supply Disruptions and Options for Mitigating their Effects,* United States General Accounting Office, Washington, DC.

GEF (1996), *Operational Strategy*, Global Environment Facility, Washington, DC.

GIPE, Paul (1995), *Wind Energy Comes of Age*, Wiley, New York.

GOLDEMBERG, José and Walter REID (1998), "Developing countries are combating climate change", *Energy Policy*, 26 (3): 233-237.

GOPALAN, H.N.B. and S. SASKENA, Eds. (1999), *Domestic Environment and Health of Women and Children*, Tata Energy Research Institute — UNEP, New Delhi.

GREENE, D.L., and P.N. LEIBY (1993), *The Social Cost to the U.S. of Monopolization of the World Oil Market, 1927-1991*, Report No. ORNL-6744, Oak Ridge National Laboratory, Oak Ridge, TN.

GROVER, P.D. (2000), "Cost Estimates for a "Dream Stove" for Asia", in FAO (2000), pp. 80-92.

GUPTA, A.K. (1999), "Biomass Energy in India – Programmes and Policies", *Wood Energy News* 14(2).

HAGLER-BAILLY (1995), *New York Environmental Externalities Cost Study*, New York.

HAITES, E. and F. MISSFELDT (2001), "Liability Rules for International Trading of Greenhouse Gas Emissions Quotas", *Climate Policy* 1 (2001), pp. 85-108.

HAMILTON, Kirk and Michael CLEMENS (1998), *Genuine Savings Rates in Developing Countries*, World Bank, Washington, DC, available at http://www-esd.worldbank.org/eei.

HAMILTON, Kirk and Ernst LUTZ (1996), *Green National Accounts: Policy Uses and Empirical Experiences*, World Bank, Washington, DC.

HICKMAN, B.G., H. G. HUNTINGTON and J.L. SWEENY, Eds. (1987), *Macro-Economic Impacts of Energy Shocks*, North Holland, Amsterdam, Netherlands.

HUNT, S. and G. SHUTTLEWORTH (1996), *Competition and Choice in Electricity*, J. Wiley & Sons, West Sussex.

IEA (1994a), *Development and Deployment of Technologies to Respond to Global Climate Change Concerns*, OECD/IEA, Paris.

IEA (1994b), *Energy Policies of IEA Countries*, OECD/IEA, Paris.

IEA (1994c), *Scoping Study: Energy and Environmental Technologies to Respond to Global Climate Change Concerns*, OECD/IEA, Paris.

IEA (1995a), *The History of the International Energy Agency*, Volume III Principal Documents, Paris.

IEA (1995b), *Natural Gas Security Study*, OECD/IEA, Paris.

IEA (1997a), *Biomass Energy: Key Issues and Priority Needs*, Conference Proceedings, OECD/IEA, Paris.

IEA (1997b), *Energy Technology Availability to Mitigate Future Greenhouse Gas Emissions*, OECD/IEA, Paris.

IEA (1997c), *Renewable Energy Policy in IEA Countries. Volume I: Overview*, OECD/IEA, Paris.

IEA (1998a), *Benign Energy? The Environmental Implications of Renewables*, OECD/IEA, Paris.

IEA (1998b), *Energy Efficiency Initiative*, Volumes I and II, OECD/IEA, Paris, available at http://www.iea.org/pubs/studies/files/danish/index.htm

IEA (1998c), *Natural Gas Pricing in Competitive Markets,* OECD/IEA, Paris.

IEA (1998d), *The Evolving Renewable Energy Market*, OECD/IEA, Paris.

IEA (1998e), *World Energy Outlook 1998*, OECD/IEA, Paris.

IEA (1999a), *Benign Energy? The Environmental Implications of Renewables*, OECD/IEA, Paris.

IEA (1999b), *Electricity Market Reform: An IEA Handbook*, OECD/IEA, Paris.

IEA (1999c), *Electricity Reform: Power Generation Costs and Investment*, OECD/IEA, Paris.

IEA (1999d), *Getting the Prices Right: Looking at Energy Subsidies*, World Energy Outlook Insights 1999, OECD/IEA, Paris.

IEA (1999e), *Regulatory Reform in Argentina's Natural Gas Sector*, OECD/IEA, Paris.

IEA (2000a), *Competition in Electricity Markets: An IEA Handbook*, OECD/IEA, Paris.

IEA (2000b), *CO_2 Emissions from Fossil Fuel Combustion – 2000 Edition*, OECD/IEA, Paris.

IEA (2000c), *Dealing with Climate Change – Policies and Measures in IEA Member Countries.* OECD/IEA, Paris.

IEA (2000d), *Electricity Information 2000*, OECD/IEA, Paris.

IEA (2000e), *Energy Policies in IEA Countries – 2000 Review*, OECD/IEA, Paris.

IEA (2000f), "Energy Subsidies in OECD Countries", EAD Working Paper, Paris.

IEA (2000g), *Energy Technology and Climate Change – A Call to Action*, OECD/IEA, Paris.

IEA (2000h), *Experience Curve for Energy Technology Policy*, OECD/IEA, Paris.

IEA (2000i), *Fuel Economy Improvement: Policies and Measures to Save Oil and Reduce CO_2 Emissions,* IEA Information Paper issued at the Sixth Conference of Parties to the United Framework Convention on Climate Change, 13 November 2000.

IEA (2000j), "Oil Prices and Gasoline Taxes: Some Economic Facts", EAD Working Paper, Paris.

IEA (2000k), *Regulatory Reform: European Gas*, OECD/IEA, Paris.

IEA (2000l), *The Road from Kyoto: Current CO_2 and Transport Policies in the IEA*, OECD/IEA, Paris.

IEA (2000m), *World Energy Outlook 2000*, OECD/IEA, Paris.

IEA (2001), "Energy Market Reform and the Environment", EDD paper, Paris.

IEA (2001 forthcoming) *Policies and Measures: Transport volume*, OECD/IEA, Paris.

IEA/UNEP (2001), "Energy Subsidy Reform and Sustainable Development: Challenges for Policymakers", Synthesis Report, Paris.

IPCC (2000), *Methodological and Technological Issues in Technology Transfer*, Special Report of the Intergovernmental Panel on Climate Change, Cambridge University Press.

JAGADEESH, A. (2000), "Wind energy development in Tamil Nadu and Andhra Pradesh, India. Institutional dynamics and barriers – A case study", *Energy Policy* 28: 157-168.

JOSKOW, Paul (1997), "Restructuring, Competition and Regulatory Reform in the US Electricity Sector", *Journal of Economic Perspectives* 11(3): 119-138.

KAMMEN, Dan (1999), "Promoting Appropriate Energy Technologies in the Developing World", *Environment* 41(5): 34-41.

KEPPLER, Jan Horst (1996), "Policy Instruments for Environmental Improvement in a Changing Market – An Overview", Proceedings of the Second EC/OECD/IEA Workshop on Energy Externalities, Brussels, 9-10 September 1996, European Commission DG XII, Brussels, pp. 51-59.

KEPPLER, Jan Horst (1998), "Externalities, Fixed Costs and Information", *Kyklos* 51:4, 547-563.

KEPPLER, Jan Horst (2001), "Energy Markets for Sustainable Development: Project Proposal", unpublished LTO paper.

KEPPLER, Jan Horst and Tom KRAM (1996), *Energy Markets — Full Cost Pricing,* Working Paper 3 for the Annex I Experts Group, OECD/IEA, Paris.

KUNTE, Arundhati, Kirk HAMILTON, John DIXON and Michael CLEMENS (1998), *Estimating National Wealth: Methodology and Results*, World Bank, Washington, DC.

LEW, D.J. (2000), "Alternatives to coal and candles: wind power in China", *Energy Policy* 28: 271-286.

LIND, Robert C. (1982), *Discounting for Time and Risk in Energy Policy*, Resources for the Future, Johns Hopkins University Press, Baltimore, MD.

LUCAS, N., T. PRICE and R. TOMPKINS (1995), "Diversity and Ignorance in Electricity Supply Investment: a Reply to Andrew Stirling", *Energy Policy*: 5-7.

LUNDVALL, Bengt-Ake (1999), "Technology policy in the learning economy" in Daniele Archibugi, Jeremy Howells and Jonathan Michie, Eds., *Innovation Policy in a Global Economy*, Cambridge University Press, pp. 19-34.

LUO, Z. and W.S. HULSCHER (2000), "Woodfuel Emissions and Cost", in FAO (2000), pp. 95-102.

MACHIAVELLI, Niccolò (1515), *Il principe*, Firenze at http://www.fauser.it/biblio/index007.htm.

MARION, N.p. and L. SVENSSON (1986), "The Terms of Trade Between Oil Importers", *Journal of International Economics* 20: 99-113.

MARTINOT, Eric (1998), "Monitoring and Evaluation of Market Development in World Bank/GEF Climate-Change Projects", World Bank, Washington, DC.

MARTINOT, Eric, Anil CABRAAL, and Subodh MATHUR (2000), "World bank solar home systems projects: Experiences and lessons learned 1993-2000", World Bank, Washington, DC.

MARTINOT, Eric and Omar MCDOOM (1999), "Promoting Energy Efficiency and Renewable Energy: GEF Climate Change Projects and Impacts", Washington, DC.

MAULL, H.W (1984), *Raw Materials, Energy and Western Security*, McMillan, London.

MCNEILL, J. R. (2000), *Something New Under the Sun: An Environmental History of the Twentieth-Century World*, Norton, New York.

METCALFE, J.S. (1995), "Technology systems and technology policy in an evolutionary framework", *Cambridge Journal of Economics* 19: 25-46.

MITCHELL, J., (1994), "What is the Energy Security Problem?", *International Journal of Global Energy Issues* 6(6): 293-300.

MOODY'S (1995), *Stranded Cost Will Threaten Credit Quality of U.S. Electrics*, Moody's Investors Services, New York.

MORK, K.A., H.T. MYSON and O. OLSEN (1989), *Macroeconomic Responses to Oil Price Increases and Decreases in Six OECD Countries*, Owen Graduate School of Management, Working Paper No. 89-12, Vanderbilt University, Nashville, TN.

NATARAJAN, I. (2000), "Social Cost and Benefit Analysis of the National Programme on Improved Chulas in India", in FAO (2000), pp. 38-42.

NEA (2000), "Nuclear Energy in a Sustainable Development Perspective", NEA/NDC/DOC(2000)5, OECD/NEA, Paris.

NEUMAYER, Eric (1999), "Global Warming: Discounting is not the Issue, but Substitutability is", *Energy Policy* 27: 33-43.

OCEAN PLANET (2000), "Oil Pollution" at http://www1.imo.org/oilweb/default.htm#un tech.

OECD (1992), *Development Assistance Manual - DAC Principles for Effective Aid*, Paris.

OECD (1994), *Effective Technology Transfer Co-operation and Capacity Building for Sustainable Development: Common Reference Paper*, OECD/GD(94)12, Paris.

OECD (1995), *Promoting Cleaner Production in Developing Countries - The Role of Development Co-operation*, Paris.

OECD (1997), *Capacity Development in Environment: Principles in Practice*, Paris.

OECD (1999a), *OECD Environmental Data – Compendium 1999*, Paris.

OECD (1999b), *Managing National Innovation Systems*, Paris.

OECD (2000a), "The Measurement of Sustainable Development", SG/SD(2000)10, Paris.

OECD (2000b), "Wealth, Values, Institutions: Trends in Government and Governance", Report of the Special Advisory Unit to the OECD Secretary General, SG/AU(2000)3, Paris.

OECD (2001), *Analytic Report on Sustainable Development: Chapter 2: Key Features and Principles*, SG/SD(2001)2, Paris.

OECD (2001 forthcoming), *Reconciling New Economies and Societies: The Role of Human and Social Capital*, Paris.

OECD/IEA (2000a), *Emission Baselines – Estimating the Unknown*, Paris.

OECD/IEA (2000b), *Kyoto Mechanisms, Monitoring and Compliance – A selection of recent OECD and IEA analyses on the Kyoto Protocol.* Paris, available at http://www.oecd.org/env/docs/cc/km-monitoring-compliance.pdf.

ORNL and RFF (1994), *Fuel Cycle Externalities*, Oak Ridge National Laboratory and Resources for the Future, Oak Ridge, TN.

PANDEY, M.R. (2000), "Health Impacts of Woodfuel Emission and How to Improve Them", in FAO (2000), pp. 2-10.

PEARCE, D.W. G.D. ATKINSON and W.R. DUBOURG (1994), "The Economics of Sustainable Development", *Annual Review of Energy and Environment* 19: 457-74.

PEARCE, David W., Kirk HAMILTON and G. ATKINSON (1996), "Measuring Sustainable Development: Progress on Indicators", *Environment and Development Economics* 1: 85-101.

PHILIBERT, Cédric (2000), "The economics of climate change and the theory of discounting", *Energy Policy* 27 (1999): 913-927.

PRÉSIDENCE FRANÇAISE DE L'UNION EUROPÉENNE (2000), "Shared Goals for Global Action: EU Proposal", mimeo, Paris.

PRIDDLE, Robert (2000), "The Future Energy Supply of Society: Options, Risks and Choices", Presentation given at the Swedish Academies' Seminar available at http://www.iea.web..

RADETZKI, Marian (2000), "Evolution of Fossil Fuel Markets in the Presence of Carbon Constraints", CONTINUE-project, unpublished paper.

RAMSEY, Frank (1928), "A mathematical theory of saving", *Economic Journal* 38: 543-559.

SCOTTISH NUCLEAR, Ed. (1994), *Diversity in UK Electricity Generation: A Portfolio Analysis of the Contribution of Nuclear,* ERM, London.

SHEPHERD, Dan (1998), "Creating a Market for Renewables: Electricity Policy Options for Developing Countries", World Bank, Washington, DC.

SINTON, J.E and D.G. FRIDLEY (2000), "What Goes Up: Recent Trends in China's Energy Consumption", *Energy Policy* 28: 671-687.

SINTON, J.E., M.D. LEVINE and W. QINGYI (1998), "Energy Efficiency in China: Accomplishments and Challenges", in *Energy Policy* 26(11): 813-829.

SMITH, K.R. (2000), "Fuel Emission, Health and Global Warming", in FAO (2000), pp. 15.

STIRLING, ANDREW (1994), "Diversity and Ignorance in Electricity Supply Investment – Addressing the Solution Rather than the Problem", *Energy Policy* 22: 195-216.

STIRLING, ANDREW (1995), "Diversity in Electricity Supply: A Response to Lucas et al.", *Energy Policy* 23: 8-11.

SUN, L. and Z. GU (1999), "Village Biomass Gas Supply System", *Wood Energy News* 13(3).

SUTHERLAND, Ronald J. (1991), "Market Barriers to Energy-Efficiency Investments", *Energy Journal* 12 (3).

TENNER, Edward (1997), *Why Things Bite Back*, Vintage, New York.

UNDP (2000), *Human Development Report*, Vienna: UNDP at http://www.undp.org/hdro.

UNDP, UNDESA and WEC (2000), *World Energy Assessment: Energy and the Challenge of Sustainability*, New York.

VAN DEN NOORD, Paul and Ann VOURC'H (1999), *Sustainable Economic Growth: Natural Resources and the Environment in Norway*, ECO/WPK(99)10, OECD, Paris.

VON DER FEHR, N. and D. HARBORD (1997), "Competition in Electricity Spot Markets. Economic Theory and International Experience", unpublished mimeo;

WATSON, Robert T., Address to the Sixth Conference of Parties to the United Framework Convention on Climate Change, 13 November 2000, available at http://www.ipcc.ch/press/sp-cop6.htm.

WEC (2000), *Energy for Tomorrow's World – Acting Now!* World Energy Council, London.

WENE, Clas-Otto, Alfred VOSS and Tanja FRIED, Eds. (2000), *Experience Curves for Policy Making – The Case of Energy Technologies*, Proceedings of the IEA International Workshop at Stuttgart, 10-11 May 1999, Institut für Energiewirtschaft und Rationelle Energieanwendung, Stuttgart.

WHO (2000), *Guidelines for Air Quality*, World Health Organisation, Geneva.

WILLIAMS, R.H. and G. TERZIAN (1993), "A benefit/cost analysis of accelerated development of photovoltaic technology", Center for Energy and Environmental Studies, Princeton.

WORLD BANK (1999a), "Bank Energy Sector Management Assistance Program, 1993-1998", ESMAP Report 223/99, Washington, DC.

WORLD BANK (1999b), *Project Appraisal Document on a Proposed Loan to the People's Republic of China for a Renewable Energy Development Project*, Report N°18479-CHA, Washington DC.

WORLD BANK (2000a), "A Summary Report from Village Power 2000 Conferences and Workshops", 4-8 December 2000, Washington, DC, *Sustainable Developments* 44(1), 15 December 2000, pp. 1-11 available at http://www.iisd.ca/sd/vp2k.

WORLD BANK (2000b), *Fuel for Thought: An environmental strategy for the energy sector*, Washington, DC available at http://www.worldbank.org/environment/pubs_data.htm.

WORLD BANK (2000c), *Toward an Environment Strategy for the World Bank Group*, Washington, DC available at http://www.worldbank.org/environment/strategy/index.shtml.

WORLD ECONOMIC FORUM, COLUMBIA UNIVERSITY and YALE UNIVERSITY (2001), *2001 Environmental Sustainability Index*, An Initiative Of The Global Leaders For Tomorrow Task Force, World Economic Forum, Yale Center for Environmental Law and Policy, New Haven available at http://www.ciesin.columbia.edu/indicators/ESI.

WWF (1995), *Delhi Environmental Status Report, An Information Handbook for Citizen Action*, Worldwide Fund for Nature, New Delhi.

YAO, X.J. (2000), "Sustainable Development of Wood Energy in China", in FAO (2000), pp. 49-55.

ZHENG, Luo (1998), "Wood Energy and the Global Environment in China", *Wood Energy News* 13(3).

Order Form

INTERNATIONAL ENERGY AGENCY

OECD

ORGANISATION FOR ECONOMIC CO-OPERATION AND DEVELOPMENT

OECD PARIS CENTRE
Tel: (+33-01) 45 24 81 67
Fax: (+33-01) 45 24 19 50
E-mail: distribution@oecd.org

OECD BONN CENTRE
Tel: (+49-228) 959 12 15
Fax: (+49-228) 959 12 18
E-mail: bonn.contact@oecd.org

OECD MEXICO CENTRE
Tel: (+52-5) 280 12 09
Fax: (+52-5) 280 04 80
E-mail: mexico.contact@oecd.org

Please send your order by mail, fax, or e-mail to your nearest IEA sales point or through the online service: www.oecd.org/bookshop

OECD TOKYO CENTRE
Tel: (+81-3) 3586 2016
Fax: (+81-3) 3584 7929
E-mail: center@oecdtokyo.org

OECD WASHINGTON CENTER
Tel: (+1-202) 785-6323
Toll-free number for orders:
(+1-800) 456-6323
Fax: (+1-202) 785-0350
E-mail: washington.contact@oecd.org

I would like to order the following publications

PUBLICATIONS	ISBN	QTY	PRICE	TOTAL
☐ **Toward a Sustainable Energy Future**	**92-64-18688-3**		**$100**	
☐ China's Worldwide Quest for Energy Security	92-64-17648-9		$100	
☐ Competition in Electricity Markets	92-64-18559-3		$75	
☐ Dealing with Climate Change: Policies and Measures in IEA Member Countries	92-64-18560-7		$100	
☐ Energy Technology and Climate Change – A Call to Action	92-64-18563-1		$75	
☐ Oil Supply Security: The Emergency Response Potential of IEA Countries in 2000	92-64-18575-5		$100	
☐ The Road from Kyoto – Current CO_2 and Transport Policies in the IEA	92-64-18561-5		$75	
☐ World Energy Outlook 2000	92-64-18513-5		$150	
			TOTAL	

DELIVERY DETAILS

Name ____________ Organisation ____________

Address ____________

Country ____________ Postcode ____________

Telephone ____________ Fax ____________

PAYMENT DETAILS

☐ I enclose a cheque payable to IEA Publications for the sum of US$ ____________ or FF ____________

☐ Please debit my credit card (tick choice). ☐ Access/Mastercard ☐ Diners ☐ VISA ☐ AMEX

Card no: ____________

Expiry date: ____________

Signature: ____________

Visit our Web site: www.iea.org

IEA PUBLICATIONS, 9, rue de la Fédération, 75739 PARIS CEDEX 15
PRINTED IN FRANCE BY STEDI
(61 01 04 1P) ISBN 92-64-18688-3 – 2001